The Korean War

Great Battles for Boys

Joe Giorello

with

Sibella Giorello

This book is dedicated to my son, Nico Giorello,
United States Marine Corps.

CONTENTS

KOREAN WAR OVERVIEW

BEFORE WE GET into the "blood and guts" of the Korean War battles, you need to know some history about the country of Korea.

Let's start in 1905.

At that time, Japan and Russia were fighting each other. These two countries are located near Korea. The conflict between them was known as the Russo-Japanese War.

Why were Russia and Japan fighting?

Russia wanted to expand its territory, and Japan wanted to halt that expansion. The battles raged for about a year before Japan won the war. The victory was unexpected because the Japanese military was much smaller than Russia's army, but the Japanese soldiers were better trained and had a more modernized navy.

As the winner of the war, Japan enjoyed the "spoils of war." The spoils are riches and resources taken by the victory from the losing country during or after the war. In this case, Japan seized control of the Korean Peninsula.

Look at the map below. Locate the countries of Russia, Japan, and China. Now check out that long stretch of land beneath Russia and China that dangles into the sea. That land is the Korean Peninsula. Notice, too, there's a line dividing the Korean Peninsula into two separate countries, North Korea and South Korea. We'll come back to that border line in a minute.

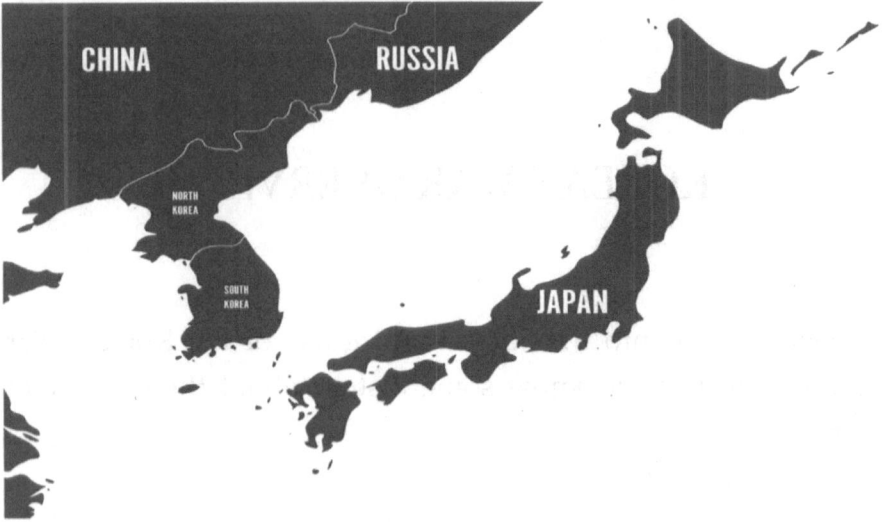

Fast-forward about forty-five years after the Russo-Japanese War, and Japan then found itself losing a war—World War II. The war's six years of continuous fighting had decimated the Japanese military and the country's economy. But despite those dire circumstances, Japan refused to surrender. The United States, which opposed Japan during WWII, sent many warnings. If Japan did not surrender immediately, it would suffer severe consequences.

Japan ignored the warnings.

In response, the US dropped two atomic bombs on Japan in August 1945. Those explosions wiped out two cities. Nothing was left but cinders and ashes.

Japan finally surrendered.

But soon after that surrender, Russia declared war on Japan. Of course, declaring war on an already-demolished country is like kicking a guy when he's down. But in war, countries don't always play fair.

Also, Russia had changed. For one, in 1917, Russia suffered a devastating revolution. Violent protestors killed the corrupt

Russian king—a leader known as the czar (pronounced "zar")—and these revolutionaries created a new kind of government. Instead of being ruled by a king, the new country grew into the Union of Soviet Socialist Republics—or USSR—ruled by Communist leaders.

Josef Stalin

Communism is a political ideology. It says the government should own everything—literally everything and divide it equally among the people. The concept is that everyone will then have the same amount, and common ownership will work for everyone's good.

Here's one example. Let's say you're a farmer. Under Communism, the government owns your land. It also owns the crops you produce on that land, and any money you earn from the sale of those crops. You *work* as a farmer, but you don't *own* any part of the farm.

The same situation is true for every other farmer in your country.

The problem is, somebody needs to be in charge of this system. That role falls to the Communist dictator—a ruler with absolute power. The dictator runs the Communist government, and he decides who gets what from your farm and from everyone else's farm. The dictator can also take as much as he wants, and no one can stop him. Under this system, the Communist dictator keeps gaining more and more power over people.

There's an old saying: "Power corrupts, and absolute power corrupts absolutely."

Soviet dictator Josef Stalin grew into a ruthless political leader. Absolute power corrupted Stalin absolutely.

In 1945, as WWII was ending, Stalin felt confident about declaring war on Japan. After WWII, the Soviets and the US (as victors in the war) took control of the Korean Peninsula from Japan. They divided Korea in half.

The Soviets would control North Korea.

The United States would control South Korea.

The line dividing these two countries would be the 38th parallel.

Check out the map below.

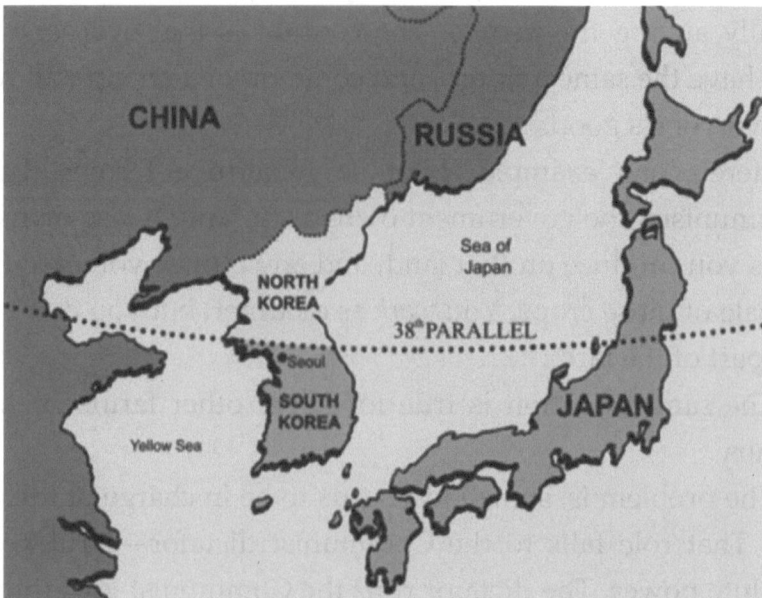

You might be wondering: What's a "parallel"?

World maps have a grid of lines that run east-to-west (left-to-right) and north-to-south (top-to-bottom).

The lines running east-to-west are latitudes, also known as "parallels." Each parallel marks the distance from the earth's equator—the middle of planet Earth. When we say the "38th parallel," it refers to a point on the globe that is 38 degrees north of the equator.

But the 38th parallel didn't solve the differing ideologies between these two countries.

Under Stalin's rule, North Korea became a full Communist country. North Korea's leader-by-title was Kim Il-Sung, a cruel dictator. He was mostly Stalin's puppet, doing whatever Stalin told him to do. Under Kim Il-Sung, the North Korean people lost many freedoms, including the right to vote. They soon fell into poverty and starvation.

Meanwhile, South Korea was headed by a democratically elected president, Syng Man Rhee. Backed by the United States, Rhee was elected by the South Korean people—who could also vote him out of office. South Korea's living conditions and freedoms were superior to those of their Communist neighbors to the north.

But a new referee had walked onto the field of battle—the United Nations. Known as the UN, the United Nations was organized after WWII to keep peace among nations and to help avoid wars, especially world wars.

The UN insisted that the North Korean people should have the right to vote. The people could decide for themselves if they wanted to live under a Communist government or if they wanted to be a democracy like South Korea.

But Josef Stalin and Kim Il-Sung refused to allow any general elections in North Korea.

United States President Harry S. Truman grew concerned. Truman worried that Communism would spread to other countries, oppressing more people. The world had just wasted too much blood and treasure defeating a group of WWII dictators such as Adolf Hitler in Germany and Benito Mussolini in Italy.

Dictatorships caused problems worldwide.

Truman, along with the UN, did not want Communism to spread. Yet China had already become a Communist country. It was ruled by a murderous dictator, Mao Tse-tung (pronounced "mow say tong").

Look at those maps again. Can you see who is Korea's nearest neighbor?

China.

For several years, there was an uneasy peace on the Korean Peninsula. This is what's called a "cold war"—no shots fired and no wars declared. Neither the Soviets nor the US wanted a "hot war." Both sides had atomic bombs, and both sides had witnessed the terrible destruction that those bombs inflicted on Japan. Further, both sides were still recovering from the ravages of WWII.

But suddenly in 1950, the cold war in Korea turned hot.

What happened?

That's what you're going to read about next—the devastating battles that erupted into the Korean War.

THE INVASION OF SOUTH KOREA

June 25, 1950

US Marines, carrying M-1 rifles and Browning Automatic Rifles, fight to liberate Seoul.

IN THE EARLY morning hours of Sunday, June 25, 1950, the people of South Korea woke to the sounds of a tremendous storm.

Flashes of lightning.

Rumbles of thunder.

But the sounds grew even louder—the thunderbolts were smashing into the ground, blowing the earth into pieces. As the explosions continued, the South Koreans suddenly realized the terrible truth.

This "storm" was an artillery barrage. Thousands of guns were firing all at once.

Thrown into panic and chaos, the South Koreans grabbed their children, their elderly family members, and their most cherished possessions—then ran for their lives.

But the artillery barrage was only the beginning of their terror.

Immediately following that fury of gunfire, about 100 Soviet-built tanks smashed through South Korea's border defenses at the 38th parallel. The monstrous tanks were followed by more than 100,000 North Korean soldiers marching forward, guns blazing.

Imagine what it was like for boys in a quiet South Korean town. In one violent moment, everything changed. Artillery fire turned their entire neighborhoods to rubble. Every sense of security suddenly evaporated in a cloud of gun smoke. Entire families were suddenly on the run, homeless, desperate to find food, safe drinking water, and safe shelter.

South Koreans flee their homes after the North Korean invasion.

The North Korean Army (we'll call them the NKA) attacked South Korea across a 200-mile wide front. The plan was to capture South Korea's capital, the city of Seoul, and force a quick surrender.

South Korea had about 100,000 men in its army. But when this attack began, many of those soldiers were on leave, taking time off from military duties. The remaining soldiers were not equipped to stop this powerful NKA attack. Also, the South Korean Army (SKA) had no tanks or even any heavy artillery weapons that could stop the Russian tanks.

In the invasion's first few hours, thousands of SKA soldiers were killed.

Hoping to salvage what remained of their county and its people, the SKA formed a fighting retreat, also known as a withdrawal. That's when an army fights while backing up. The SKA tried other defensive tactics, too, such as trench warfare. Soldiers dug channels into the ground and hunkered inside with their weapons. But the NKA machine guns mowed them down and the Soviet tanks—known as T-34s—barreled forward with cannons belching flames of fire.

The SKA's only effective weapons against these tank assaults were hand grenades and light bazookas. But these exploding projectiles sometimes bounced off the fire-breathing tanks. Some SKA soldiers bundled hand grenades together, trying to create a more effective weapon. Still others, in desperation to save their country, sacrificed their lives by throwing their own bodies onto the tanks.

The slaughter lasted for days.

Finally, the SKA managed to put up a strong defensive line at the Han River, holding back the enemy and blocking them from moving forward. That defense line also ruined the plan of

North Korean dictator Kim Il-Sung, who wanted to quickly capture Seoul and force South Korea to surrender.

UN forces fighting on the outskirts of Seoul, 1950

News about this catastrophic invasion spread around the world. The United States—as the "caretakers" of South Korea—went to the United Nations and asked for military support to fight the NKA. The UN approved the request, and US President Harry S. Truman agreed to send American forces into the conflict. (Side note: Truman's order was the first time in US history that a sitting president sent troops into an international conflict without first asking Congress for a declaration of war.)

On July 1, 1950, the first US military forces arrived in Pusan, South Korea. Pusan is a port—a waterfront city with a harbor for ships. Ports are strategically crucial during warfare because they help expedite armies and supplies to the battlefront.

US soldiers, dug into the hillside, firing on Communist-held NKA positions.

These first US forces, about 400 men, were part of the Army's 24th Infantry Division. Most of these soldiers were young, inexperienced, and had limited training. But these American servicemen had one crucial trait that would both help them and hurt them—pride.

Five years before this invasion, the United States military had helped win World War II. Given the fresh memory of that amazing victory, the US forces were convinced that once the NKA faced the formidable US military, they would flee.

That didn't happen.

On July 5, 1950, this portion of soldiers from the 24th Infantry Division faced the advancing NKA. From their defensive line, these 400 soldiers fired guns, hurled grenades, and hollered to rally their fellow men through the smoke and noise.

But rather than run away from this force, the NKA charged

from all sides. The Soviet-made T-34 tanks rolled forward, blasting fire. It quickly became evident that these first US forces were nearly as ill-equipped to fight back as the SKA—they had arrived without the necessary weapons to stop the Soviet tanks.

The US soldiers went into a full retreat.

United Nations soldier fires a submachine gun on Communist-led forces in Seoul.

Fortunately, help was coming. The rest of the 24th Infantry Division—about 16,000 men—arrived in South Korea under the command of Major General William Dean.

UN gun crew near the Kum River, South Korea, July 15, 1950.

Dean placed his men into a new defensive position. But, once again, without heavy weapons to combat the onslaught, they were unable to stop the NKA from pushing forward.

The 24th fell into another retreat.

Even worse, Dean was captured in the retreat! It was a devastating blow for the US forces—their leader was now a prisoner of the enemy.

Morale sank among the US forces.

But reinforcements arrived, mostly from the US and Great Britain. Known as the 8th Army, these men were led by Lieutenant General Walton Walker. Perhaps even better, US and British tanks also arrived to combat the Soviet-made T-34s.

This fresh infusion helped the UN forces hold a perimeter, or boundary, around the port city of Pusan.

Pershing and Sherman tanks of the 73rd Heavy Tank Battalion at the Pusan Docks.

On July 7, more help arrived. US General Douglas MacArthur was named the Commander in Chief, Far East Command.

MacArthur, a legendary military general, was placed in command of all soldiers from every country that was coming to help South Korea—and many countries wanted to halt this violent expansion of Communism. In addition to the United States, Britain, Canada, and Australia, dozens more countries sent aid in the form of soldiers or equipment.

MacArthur realized that given the dire situation in South Korea, a powerful counterattack was needed.

But what kind of counterattack?

That's what you're going to read about in the next chapter.

WHO FOUGHT

General Douglas MacArthur with his corncob pipe.

DOUGLAS MACARTHUR WAS born on January 26, 1880.

His father, Arthur—yes, his name was Arthur MacArthur!—was a US Army captain who had fought for the Union during the American Civil War.

When Douglas was thirteen years old, the MacArthur family moved to San Antonio, Texas, the site of the famous American battle The Alamo. Douglas attended the West Texas Military Academy, then enrolled in the United States Military Academy at West Point. In 1903 Douglas graduated with honors and began his rise through the US Army's ranks.

MacArthur earned a strong battlefield reputation in World War I. After Japan bombed Pearl Harbor in 1941, propelling America into World War II, MacArthur was stationed in the

Philippine Islands. Japan soon attacked there and MacArthur countered with several successful offensive operations. However, MacArthur was a strong-willed man and an outspoken soldier. He was frequently—and openly—critical of the decisions made by his superiors. Because of that, President Truman eventually fired him, removing him from leadership in the Korean War.

MacArthur was respected as much as he was despised—men either loved him or hated him. Fearless on the battlefield, MacArthur uttered many memorable statements:

"It is fatal to enter a war without the will to win it."

"Age wrinkles the body; quitting wrinkles the soul."

"We are not retreating—we are advancing in another direction."

Despite all his controversy, when General Douglas MacArthur died, he was honored with a state funeral and was buried at the MacArthur Memorial in Norfolk, Virginia. The memorial also houses a museum collection that documents the life of this passionate American whose dedicated military service changed the world.

BOOKS

The Start and End of the Korean War – History Book of Facts by Baby Professor

America In Korean War: A History Just for Kids! by KidCaps

The Korean War: An Interactive Modern History Adventure by Michael Burgan

INTERNET

The Korean War, by Simple History:
youtube.com/watch?v=yxaegqvl4aE

General Douglas MacArthur's life and military service:
youtube.com/watch?v=JJb5bdelws4

Fourteen interesting photos of the Korean War, presented by
History.com: history.com/topics/korea/korean-war#section_6

MOVIES

One Minute to Zero (1952), starring actor Robert Mitchum. You can
watch the movie's trailer here:
youtube.com/watch?v=rwJV5EdRk60

THE BATTLE OF INCHON

September 15, 1950

USS *Toledo*'s eight-inch guns fire on military targets during the Korean War, 1950.

HAVE YOU EVER heard of a "sucker punch"?

That's a punch thrown without warning. And because the victim isn't prepared, a sucker punch usually knocks him to the ground and leaves him vulnerable to the next blow.

With its surprise invasion on South Korea, the NKA had

thrown a military sucker punch.

But General Douglas MacArthur decided to throw his own sucker punch in return.

MacArthur planned to invade exactly where nobody expected UN forces to land—150 miles *behind* enemy lines.

For his invasion, MacArthur chose the port city of Inchon, which is on the west coast of the Korean Peninsula.

"We shall land at Inchon," MacArthur said, "and I shall crush them."

MacArthur's plan was extremely risky. To this day, historians disagree about whether the Inchon landings—code-named Operation Chromite—were a good idea.

Here are just ten reasons why these landings could have failed.

General Douglas MacArthur (center) aboard the USS *Mount McKinley* during the shelling of Inchon.

Top Ten Risks of the Inchon Landings

ONE, IN ORDER to reach Inchon, the military task force would have to navigate thirty miles through a narrow and hazardous channel.

Two, the channel was littered with explosive mines that could destroy the ships. The mines might also damage a ship so badly that it blocked the channel, preventing the rest of the invasion force from completing the mission.

Three, the NKA set up shore batteries. These cannons would fire down on the incoming invasion force.

Four, the channel's ocean currents were so powerful they might push the landing vessels off course.

Five, the channel's high and low tides were extreme, rising and falling by as much as thirty feet. In order to deliver the men to shore, the landing crafts needed to be timed with a high tide.

Six, if the landing happened at low tide, the mud flats would act like quicksand, trapping the men in place.

Seven, Inchon didn't have sandy beaches. It was a rocky coastline fortified with a concrete seawall. Even if the landing force managed to reach shore, the men would face even more obstacles ahead.

Eight, the men would need to climb over that seawall, exposing them to enemy fire.

Nine, a small island named Wolmi-Do sat in front of the Inchon coastline. Enemy soldiers controlled the island and would need to be defeated before any invasion force could reach Inchon.

Ten, the mission's ultimate goal was to retake control of the city of Seoul. But Seoul was more than twenty miles from Inchon. If the landing force did succeed at Inchon, it still had a long fight ahead to reach its final destination.

A battleship fires on Korea to give cover for men on land.

MacArthur's invasion plan included hundreds of warships from the US, Australia, Canada, New Zealand, France, Netherlands, and Great Britain. As the men fought their way to shore, these ships would provide cover through artillery power. In total, this landing force would number about 70,000 men known as the X Corps or the Tenth Corps ("corps" is pronounced "core"). This force included the 1st Marine Division and the Army's 7th Infantry Division.

On September 10, ships and planes began a bombardment. The explosives were fired at the island of Wolmi-Do. Three days later, two US heavy cruiser ships, two British light cruisers, and six US destroyers demolished enemy gun placements along the Korean shore. That action allowed a large force of Marines to progress forward. But nobody knew what those men would face.

"We'll find out what's on the beach when we get there," Marine Colonel Lewis "Chesty" Puller told his men. "We live by the sword, and if necessary, we'll be ready to die by one. Good luck. I'll see you ashore."

Imagine this scene:

It's early morning on September 15. You scale the sides of an enormous ship by climbing down a netted-rope wall that sways back and forth. At the bottom of the rope wall, you leap into a small landing craft that bobs up and down in turbulent water. You wait in the landing craft until it's full of men, then the vessel motors in circles around the larger ship, waiting for the other landing crafts to fill. You glance at the guy next to you. Sweat beads across his face. Other men have a distant stare. Some whisper prayers. But all of you show courage by remaining in place despite feeling terrified inside. Then you show even more courage as the landing crafts speed toward their military objective—sending you directly into danger.

South Korean Marines descend into the landing craft that carried them to Inchon.
US Army photo

The Marines were given the first objective on Wolmi-Do island. Their objective was a location called Green Beach. This initial Marine landing force was led by Lieutenant Colonel Robert Taplett.

At first, the Marines didn't experience heavy enemy fire. But as soon as their nine M26 Pershing tanks came ashore, enemy guns opened up.

The Marines hurried ashore under deadly fire. One tank

armed with a flamethrower torched its way forward. Other tanks equipped with bulldozer blades plowed up the earth and buried the enemy inside their own caves.

The Marines' objective on Wolmi-Do was to take the island's highest point, a place called Radio Hill. After 47 minutes of hard fighting, Taplett and his men seized control of the island. The Marines then raised the US flag on the hill. In total, they had incurred twenty casualties while killing more than 100 NKA soldiers and capturing another 136.

Enemy soldiers surrender to US Marines during attack on Wolmi-Do.

Next the mission moved to Inchon.

But remember those extreme ocean tides? The Marines on Wolmi-Do needed to wait until 5:30 p.m. for the next high tide that would raise the waterline high enough for ships to reach the Inchon shore. This waiting game triggered a new concern— would the North Korean enemy send reinforcements to Wolmi-

Do to surround the US Marines and possibly wipe them out?

Fighter planes—Corsairs and Skyraiders—circled the airspace over the island to protect the Marines.

Then at 2:30 p.m., UN warships began bombarding Inchon. Fire blazed from the ships' powerful guns that were aimed not only at the landing area, but also twenty-five miles behind it in order to block any NKA reinforcements.

Two main landing forces were designated for Inchon. The 5th Marines would take an area called Red Beach. The 1st Marines would arrive at Blue Beach. Look at the map below to see these beaches. Also, notice the location of Wolmi-Do and Inchon.

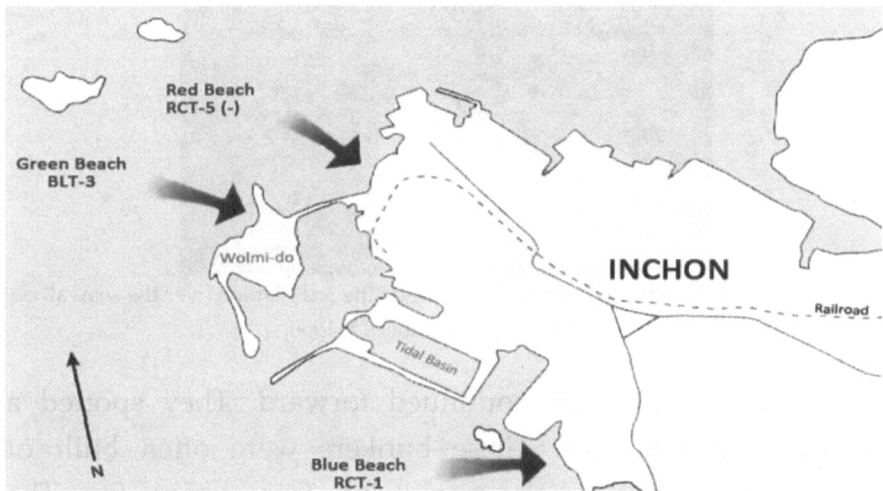

When the 5th Marines landed, they faced a 1,000-foot-long seawall. In some locations this wall rose more than fifteen feet. To scale the wall, the Marines used ladders and scrambled up under enemy fire.

First Lieutenant Baldomero Lopez was leading the men of the 3rd Platoon. Lieutenant Lopez was the first Marine over the wall. Immediately he was confronted by twelve North Korean soldiers. Lopez lowered his weapon and fired, taking out all twelve enemies.

First Lieutenant Baldomero Lopez, USMC, leads the 3rd Platoon over the seawall on Red Beach, September 15, 1950.

Lopez and his men continued forward. They spotted a North Korean bunker. These bunkers were often built of concrete that protected the men inside from enemy fire. The only way to destroy the bunkers was to get really close and hit it very hard. Lopez and his men did just that—they wiped out the bunker and the men inside.

But the Marines then spotted another bunker.

Lieutenant Tom Gibson was among Lopez's men.

"Lopez began to attack it," Gibson said. "Before he could throw the grenade he held in his hand, he was hit. The grenade dropped to his side. To save the men of his platoon he rolled over on top of it."

Lopez saved his men—but was killed during the action.

More men from the 5th Marines scaled the seawall to join the fight. Fortunately, naval bombardment had blasted holes in the seawall, making it easier to climb over. But enemy gunfire grew more intense.

The Marines pressed forward. They were heading for their objective, a high point called Cemetery Hill. Using all their firepower, the Marines forced the North Koreans to break and run. But the Marines did not let up. They fought their way into the streets of Inchon. Within twenty minutes, a flare shot into the sky. It was a signal to the ships that the Marines had succeeded in their mission.

But what about the 1st Marines at Blue Beach?

They faced more seawall. Some men used dynamite to blow openings in the wall. But it was getting late and darkness was closing in around them. Further complicating matters, all the smoke from the battle combined with the evening fog off the water made it difficult for the incoming landing craft to see where to drop off the Marines. Some of the first crafts came in short of their target. To reach the beach, men were forced to wade through icy cold ocean water under heavy enemy fire. These Marines also wore heavy packs that turned them into slow-moving targets for the NKA.

But the Marines never relented. They eventually took control of the operation and successfully landed on Inchon.

By 1:30 a.m. the next day, the city of Inchon was surrounded—and under UN control.

Ten days later, MacArthur's force freed the city of Seoul, twenty miles away. The Marines and the Army's 7th Infantry Division also gained control of Kempo Air Field.

American M26 Pershing tanks in downtown Seoul as UN troops round up North Korean prisoners of war.

Even farther away, at the port city of Pusan, the UN forces launched an offensive, attacking and driving the NKA out of what was known as the "Pusan perimeter." On September 26, the Marine, Army, and UN forces linked up and destroyed eight North Korean divisions. Any NKA survivors raced for the 38th parallel, which still divided North and South Korea.

Perhaps the whole Korean conflict might have ended right there. After all, the situation was right back where it began with a border at the 38th parallel.

But the victorious feeling among US and UN leaders poisoned the atmosphere. One historian later described it as "victory disease" or a "false sense of optimism about the course of the war."

Instead of intimidating the Communists, the Inchon landings and the recapture of Seoul only infuriated them further. Now they planned to fight back even harder, escalating the war to new destructive heights. This decision spelled disaster for the US and UN forces.

The Korean War would now become more devastating than anyone imagined.

WHO FOUGHT

Baldomero Lopez

BALDOMERO LOPEZ WAS born August 23, 1925, in Tampa, Florida. In high school he was a star basketball player and regimental commander of his school's Junior Reserve Officers' Training Corps (JROTC) program. In 1943, during WWII, he enlisted in the Navy and served for one year before being appointed to the US Naval Academy.

As a second lieutenant in the Marine Corps, Lopez was sent to China where he served as a rifle platoon commander. When the Korean War broke out, Lopez volunteered for duty as an infantry officer.

On September 15, 1950, during the invasion of Inchon, Lopez was clearing enemy beach defenses when he lifted his arm to throw a grenade. Enemy automatic weapon fire struck his right shoulder and chest, causing him to drop the live grenade. Despite tremendous pain and his body hemorrhaging blood, Lopez dragged himself to the grenade. He tried to throw it again but his fingers were unable to grasp it tightly enough. In a moment of exceptional courage, Lopez swept his wounded right arm under his body, cradling the grenade and absorbing the full impact of the grenade's explosion. Lopez's valor—which is courage in the face of danger—saved the lives of the men around him.

News of Lopez's death spread quickly. One war correspondent wrote that Lopez "died with the courage that makes men great."

Posthumously—after his death—Lopez was promoted to First Lieutenant. He received many honors, including the Purple Heart for injuries sustained in battle and the Medal of Honor, the highest military award in the United States, for risking his life above and beyond the call of duty.

Several buildings were also named for First Lieutenant Baldomero Lopez, including an elementary school in Seffner, Florida, and a public swimming pool in Tampa, Florida.

BOOKS

Inch'on 1950: The last great amphibious assault by Gordon L. Rottman and Peter Dennis

INTERNET

A short YouTube video on US forces liberating Seoul, South Korea: youtube.com/watch?v=5y4h-iHSmOU

MOVIES

Inchon (1981)

THE SECOND BATTLE OF NAKTONG RIVER

September 23, 1950

UN forces take defensive positions behind the Naktong River.
Harry S. Truman Library

UP UNTIL THIS point in the war, the United States was the only solid military force defending South Korea.

But help was on the way.

The British were coming to Korea.

In August 1950, Brigadier Basil Aubrey Coad landed with the 27th Brigade at the port of Pusan. This brigade would be under the command of the US 8th Army. Their first assignment would be to close the gap along the southern flank of the longest river in South Korea, the Naktong River. In particular, cover was needed for a ten-mile front along the Naktong River.

Two forces of the 27th Brigade were given this assignment. They were part of the Middlesex Regiment, who were English, and the Argylls, who were Scottish.

The Argylls' assignment was to secure the high ground of multiple elevated positions. One rise was named Hill 282. Another hill was called Plum Pudding Hill. And a third, 900 feet high, was known as Middlesex Hill.

You might think that since this conflict was modern warfare—with jet fighters and machine guns—a bayonet charge would be a thing of the past.

Nope.

The Middlesex Regiment charged Plum Pudding Hill with their bayonets—steel blades fixed in the slot on their rifle. As the soldiers topped the rise, they stabbed their blades into enemy soldiers.

The North Koreans broke and ran down the hill.

Plum Pudding was secure.

Next was Middlesex Hill. Two US tanks joined the fight, opening fire until the hill was cleared of the enemy.

The Argylls in Korea, wearing their distinctive Scottish tam-o'-shanter hats.

Hill 282 was next, and it was the Argylls' turn. Three companies were available. Two of those companies were ordered to assault the hill. The third company was kept in reserve to provide fire support—while the other two companies moved up the hill, this third company would keep firing weapons, forcing the enemy to keep their heads down and thus be unable to fire back.

On September 23, around 5:20 a.m., the Argylls moved out and began climbing the steep slopes. The men tried to move stealthily—unseen and unheard—but loose rocks slid beneath their boots and rattled down the hill. Finally, the Argylls got close enough that they could smell the North Koreans cooking breakfast. The Argylls charged forward and blasted the enemy with rifle fire, hand grenades, and flashing bayonets.

They soon controlled that portion of the hill.

Soldier along the Naktong River fires on Communist-led North Koreans,
August 1950.

Unfortunately, the North Koreans held another high position above them. The Argylls dug in, preparing to defend their new position while expecting a counterattack by the enemy. Sure enough, the North Koreans soon opened fire with a mortar barrage. The shelling killed four Argylls and wounded nine more.

The Argylls realized this barrage was only the beginning of the counterattack. The North Koreans would only increase their firing. The Argylls called for help and received two American artillery observers.

Artillery observers were crucial in battle because they could help pinpoint an enemy's location. They would relay that information to Allied firepower who could rain destruction on them.

Unfortunately, at 8:45 a.m., these two American observers received orders to leave the hill.

Now the Argylls would have to fend for themselves.

The North Koreans launched more mortar fire. And with the artillery observers gone, the enemy infantry could now move forward undetected. Also, because this hillside was covered with thick vegetation, the North Koreans were protected by natural cover.

Closer and closer they came, picking off the Argylls until they reduced the platoon to only seven men.

Another platoon came forward to reinforce. But it took on heavy casualties, too.

Realizing his Argyll forces were about to be overrun, Second Lieutenant J.R.R. Edington, who was himself wounded, regrouped his men. The new goal was to hold the perimeter.

Machine-gun squad of the 1st Cavalry Division fires on North Korean patrols along the Naktong River, Korea.

The North Koreans kept moving forward. The Argylls' situation grew even more desperate. Finally, the order was given: "To the last man and the last round."

In other words, fight to the finish.

But the Argylls faced another problem. They were 900 feet up on the hillside and needed to get the wounded to safety. However, that task entailed retracing their slippery path down the slope—while carrying bodies and dodging enemy fire.

This difficult task was given to Sergeant Major Tom Collett. He estimated the journey down would require at least one hour, then another hour to return to the high position. That trek would help bring the wounded men to safety, but it would also remove desperately needed fighters from the Argylls' defensive position.

Collett's men feared him as much as they feared the enemy. They hustled to complete his orders. And Collett even timed their journey with a stopwatch as they raced down and up the slope. "No stopping at the bottom of the hill and lighting a [cigarette]," he said.

Brigade Church Parade, Sunday, 18th March 1951

Argylls in full dress uniforms—kilts, tams, and bagpipes—in Korea, 1951.

Fortunately, help soon arrived in the form of Major Kenneth Muir. He brought stretcher-bearers who helped carry the wounded, allowing the other men to stay in their fighting positions.

The battle raged all day. More and more men were getting wounded. But the Argylls were not winning any ground.

Major Muir reorganized the remaining soldiers from the two Argylls' companies into one company. He also streamlined communication, concentrated firepower, and reorganized ammunition and resupply lines—all while evacuating the wounded from the hillside.

As the battle stormed on, Muir's commanding presence brought morale back into the fight.

But a large force of North Koreans prepared to attack again. And the Argylls had no artillery support available. Muir asked for air strike support and was told that three P51 Mustangs fighter planes would soon arrive.

But the Mustangs would need to find Muir's position in the thick foliage. Muir laid out white panels large enough to be seen from the air that showed the pilots which terrain belonged to the Argylls and which belonged to the enemy. But the North Koreans spotted the white panels—and laid out their own white panels to confuse the pilots!

Just after noon, the planes zoomed overhead. But in the confusion of all these white panels, the napalm fire bombs—a flammable gel that incinerates anything it touches—were dropped on the enemy *and* on the Argylls.

Within two minutes, the hill erupted in flames.

When the smoke finally cleared, most of the Argylls had survived. But the men were upset. Their own side had attacked them! And now, due to the confusion, they were below their original position.

Army commanders sent the word to end this mission.

But as the Argylls began withdrawing, the men heard sudden gunfire—a Bren machine gun. It was still at the top of the hill, and someone from their side was firing it!

Muir gathered about thirty men. He refused to leave anyone behind.

"Right," he said. "We're going in."

At a blistering pace, Muir led his men up the hill while the enemy fired on them. When they reached the top, they found Private William Watts and four other men. Watts said nobody told him to leave his position—and he still had two magazines for his Bren gun. Talk about obeying orders!

But the North Koreans were far from finished with this fight. They closed in on the Argylls from three sides. Muir fired his Sten gun—a submachine gun—until he ran out of ammunition. He then grabbed a two-inch mortar, a metal tube that shoots projectiles. But while Major Alastair Gordon-Ingram was loading the mortar, Muir was hit in the stomach and thigh by enemy fire. Even as Muir was carried down the hill, he shouted for his men to keep up the fight. Unfortunately, Muir died before reaching the bottom of the hill. He was later awarded the Victoria Cross, Britain's highest military award for valor "in the presence of the enemy."

Major Gordon-Ingram stepped up to command the Argylls on the hill. Ten men remained. And some of them were wounded. The Argylls continued to fight the enemy until, finally, word came from command: "Get out of there!"

The Argylls, staggering from exhaustion, bleeding from wounds, withdrew from the hill firing the last of their ammunition.

Their survival was a small miracle—and proved to everyone that these Scottish soldiers came equipped with steel spines. Even when their destruction looked certain, the Argylls did not waiver.

As news spread about the Argylls' brave stand spread around the world, people were amazed. One survivor of the battle agreed.

"Hardly believable now," he later said, "but it was amazing what a human being can put up with when he must."

WHO FOUGHT

British Major Kenneth Muir

MAJOR KENNETH MUIR'S determination and personal courage inspired his men. His heroic actions on Hill 282 are now military legend. Instead of giving in to the demoralized mood of his men after that mistaken air strike, Muir redirected them into a counterattack. Soon the Argylls had retaken the crest of the hill—after it seemed like all was lost.

The *London Gazette* reported on the Argylls' stand.

"From this moment on," the paper said, "Major Muir's actions were beyond all possible praise. He was determined the wounded would have adequate time to be taken out and he was just as determined that the enemy would not take the crest. Grossly outnumbered and under heavy automatic fire, Major Muir moved about his small force, redistributing fast diminishing ammunition, and when the ammunition for his own weapon was spent he took over a 2-inch mortar, which he used with very great effect against the enemy. While firing the mortar he was still shouting encouragements and advice to his men, and for a further five minutes the enemy were held. Finally, Major Muir was hit with two bursts of automatic fire which mortally wounded him."

Yet even as Muir's life was fading, he remained determined.

His last words were: "They will never drive the Argylls off this hill."

"The effect of his splendid leadership on the men was nothing short of amazing," the report continued, "and it was entirely due to his magnificent courage and example and the spirit which he imbued in those about him that all the wounded were evacuated from the hill, and, as was subsequently discovered, very heavy casualties inflicted on the enemy in defence of the crest."

BOOKS

Argyll and Sutherland Highlanders by Alastair Campbell

Did you know that the Scots have never won a battle in which they were favored to win? The Scots succeeded when they were the underdog. Learn more in *Scottish History: A Captivating Guide to the History of Scotland*, part of the Captivating History series.

INTERNET

Learn more about the amazing fighting forces of the Royal Regiment of Scotland: army.mod.uk/who-we-are/corps-regiments-and-units/infantry/royal-regiment-of-scotland

MOVIES

Braveheart tells the tale of Scotland's most famous warrior, William Wallace, and his fight against the King of England.

BATTLE OF CHOSIN RESERVOIR

November 27 – December 13, 1950

A column of troops and armor of the 1st Marine Division move through Communist Chinese lines.

JUST AS THE Communists decided to escalate this war after being pushed back to the 38th parallel, General Douglas MacArthur received an order from the US Joint Chiefs of Staff. The joint chiefs are the US president's military consultants who represent the Army, Navy, and Air Force.

The chiefs told MacArthur: "Your military objective is the destruction of the North Korean armed forces. In obtaining this objective, you are authorized to conduct military operations north of the 38th parallel in Korea."

In other words, they gave MacArthur permission to wipe out the enemy—even if that meant invading North Korea. This directive would also help lead the US Marines into one of the most brutal and heroic fights of modern-day warfare: the Battle of Chosin Reservoir (pronounced "cho-sun").

This legendary battle lasted only seventeen days, but it caused the deaths of thousands of American Marines and US Army personnel. The battle also produced heroes who somehow survived near-impossible conditions while also destroying a numerically superior enemy.

The men who survived this battle were later called "the Chosin Few."

Here's what happened.

In late November, as Korea's harsh winter weather plunged to temperatures far below zero, General MacArthur ordered an attack across the 38th parallel. The men in this attack would include the US 8th Army and the Tenth Corps, which was a mix of US Marines, British soldiers, and US Army.

At this point in the Korean War, China had not formally entered the conflict. However in secret, the Chinese Communists were already moving military forces into North Korea.

On November 24, Lieutenant General Walton H. Walker sent out his 8th Army on an offensive. Walker didn't realize how many Communist Chinese soldiers were already in this fight.

The next day, the Chinese army counterattacked and sent the 8th into full retreat.

Two days later, as ice and snow gripped the frozen ground, General Edward "Ned" Almond, in charge of the Tenth Corps, ordered another offense. Almond's goal was twofold. He wanted to take some pressure off the 8th Army, and he wanted to push forward to China's border with Korea.

That second objective would place his men near the Chosin Reservoir. (A reservoir is a man-made lake that stores water).

US Marines near the Chosin Reservoir.

Three major battles—and many smaller ones, or skirmishes—were fought near the Chosin Reservoir: Yudam-ni, Hagaru-Ri, and Koto-Ri.

On November 27, the Marines at Yudam-ni heard some odd noises. At first they couldn't identify the sounds. But the rustle and crunching noises continued through the night. And then suddenly the icy air fractured. Bugles blared. Whistles shrieked. Men hollered battle cries.

The Chinese soldiers had crept up close to the Marines during the night—so close one Marine was bayonetted in his sleeping bag.

Lieutenant John Yancey yelled the order to his men: "Bayonets on rifles!"

Sliding their knives into their rifle barrels, Yancey's men

prepared for hand-to-hand combat when suddenly a Chinese soldier popped up from a foxhole. Nearby, a Marine sergeant pulled his .45 caliber handgun and shot the enemy in the face.

Then all the Marines opened fire.

But the Marines struggled to see their enemy. The Chinese soldiers wore white uniforms which worked like camouflage in the snow and ice. The Marines had to resort to firing back at the bright flashes bursting from the Chinese guns.

During the firefight, Lieutenant Yancey constantly moved from foxhole to foxhole, making sure his men were supplied with enough ammunition. Suddenly a bullet ripped into Yancey's right cheek. The bullet then lodged inside his nose.

"A spent bullet," Yancey later said.

A spent bullet is one in which the gunpowder has been expelled, leaving just the metal portion.

"Blood was oozing down my cheek into my mouth, but then it froze up. I didn't say anything about this to anyone."

This fierce attack turned out to be a probe—a scouting mission. The Chinese wanted to know the locations of the Marines' machine-gun positions. When darkness fell, the Chinese moved in for another attack, once again hoping to catch the Marines off guard. But the Americans recognized the sounds of crunching ice and snow.

"It had been quiet for an hour or two," Yancey said, "then we began hearing these odd noises. Down at the bottom of the slope, like hundreds of feet, walking slowly across a big carpet of cornflakes. They're coming up the hill."

The Marines struck back.

To reveal the enemy's location in the darkness, the Marines fired star shells. The explosions lit up the night sky just before the Marines unleashed machine gun, rifle, and mortar fire. The

Chinese countered by throwing hand grenades bunched together in clusters for greater impact.

This was a brutal and bloody fight.

But the Marines faced another enemy—the freezing cold weather. Drinking water froze inside metal canteens. Food turned to solid ice. Rifle mechanisms jammed. Frostbite disfigured fingers, toes, and faces.

A wounded chaplain reads a memorial service over Marines killed in action, Korea, 1950.

The cold also hindered care for the wounded. One medic described the challenges.

"We put as many wounded as we could in sleeping bags to keep them warm so that shock wouldn't kill them." Shock is a deadly condition in which an injured body doesn't get enough blood flow. "We tried to save our morphine [a powerful painkiller] for the marines hit in the chest or gut. You had to hold syrettes [medicine vials] in your mouth to keep them thawed out."

Through it all, Lieutenant Yancey continued his patrol from foxhole to foxhole. The bullet was still lodged in his nose— when suddenly a nearby explosion shot shrapnel into roof of his mouth!

"After that, blood kept trickling down my throat," he said. "I kept spitting it out."

On November 28, the Chinese attacked the location of Hagaru-Ri and took the high ground. That meant they could fire down on the Marines. The next day, Marine Corsair fighter planes flew in and opened fire, annihilating enemy forces on what was known as East Hill.

But as soon as the air strikes stopped, more waves of Chinese soldiers rushed in. In fact, it seemed like the Chinese had an unlimited supply of men. Later figures would estimate the Chinese had about 120,000 soldiers in the Battle of Chosin Reservoir. The UN forces were only about 30,000.

The Marines, along with Army forces, tried to retake East Hill. But they did not have enough manpower to succeed. General O.P. Smith decided to reinforce a lower perimeter around the hill. Already Smith had lost 500 men in the last 24 hours.

The least-attacked area at this point was under the command of Lewis "Chesty" Puller. But Puller suspected Hagaru-Ri was about to be overrun. He sent a relief force under the command of British Marine Lieutenant Colonel Douglas Drysdale. This 1,000-man force was known as Task Force Drysdale.

Unfortunately, the Chinese attacked the task force just as it was on its way to help. Only about 300 men survived that attack. They continued on to reinforce the men at Hagaru-Ri.

Meanwhile, also at Hagaru-Ri, Marine combat engineers managed an amazing feat. They built an airstrip—in freezing temperatures while under enemy fire. The airstrip helped evacuate more than 4,000 casualties.

Combat engineers sanding down an icy road.

Tragically, due to lack of space on the planes, some of the dead were left behind. They also couldn't be buried because the ground was frozen solid. The Marines had to resort to using a large trench that had been blasted out as an artillery position, burying their dead in a mass grave.

Amid all these challenges, the tough Marines held their defensive positions.

But the Chinese were determined to wipe them out.

On November 30, General Almond realized the situation

was dire. This mission had little chance of succeeding.

Almond ordered a withdrawal of the Tenth Corps from the Chosin Reservoir.

Specifically, Almond told General O.P. Smith to escape from the reservoir area and get to the Korean coastline as quickly as possible.

Escaping seemed impossible. The Chinese not only out-numbered these men, they also had them surrounded in a treacherous landscape of steep snow-covered mountains. Even if the Marines managed to get out alive, the coastline was another eighty miles away. Every step of that trek would be a fight.

To gain some speed, Almond advised Smith leave behind all the heavy weapons and equipment.

Smith declined.

"No sir," Smith replied. "We'll fight our way out as Marines, bringing all our weapons and gear with us."

Heavy Tank Company, 7th Infantry Division, at the Chosin Reservoir provides cover as infantrymen escort a prisoner to the rear.

Under heavy fire from the enemy, the Marines began withdrawing. But making matters even more treacherous, they could only escape on one road—the only route out of the Chosin Reservoir area. If they hoped to ever get out alive, the Marines would need to take the high ground and clear it of the Chinese forces who were firing down at them or blocking their escape at every turn.

The 1st Marine Air Wing flew in to help clear some of the steep areas while a convoy of men and machines—as many 1,000 vehicles—marched down a frozen one-lane road that snaked between the icy towering mountains that were filled with murderous enemies. The Chinese had also set up at least eight main roadblocks to pen the Americans in the valley.

But the Marines never relented, fighting for every inch of ground forward, hoping for the safety of the coastline. What they met on this trek was beyond their imaginations.

The Marines fighting at a place called Toktong Pass still held their ground but had lost about half their men. As this convoy of escaping Marines approached the pass, they saw the ground littered with hundreds of dead Chinese soldiers. Once again, outnumbered Marines had slaughtered their enemy.

At least two more major obstacles remained for these men retreating from Chosin. One was Funchilin Pass. The other was an enemy strongpoint called Bill Hill—a ridge about 1,000 feet above the ground.

At Funchilin Pass, the Chinese had blown up a bridge. The convoy stalled, unable to move forward. But the Air Force flew in and dropped eight sections of bridge works. Each of these bridge works weighed about 2,500 pounds—or the weight of five modern automobiles. Imagine dropping that from a plane! The Marine combat engineers rose to the challenge. They completed the bridge, once again under enemy fire.

Combat engineers building a floating bridge during the Korean War.

To get past Bill Hill, the Marines scaled the 1,000-foot ridge, fighting their way through snow and ice to the top where they attacked the enemy from every direction. They blew up the bunkers and other strongpoints in a multi-front attack that confused the Chinese forces. The Marines drove them off Bill Hill. Now they could protect the convoy below.

The battered forces reached the coastline on December 11.

The Battle of Chosin Reservoir was over. It was an incredible fighting retreat.

But the cost was high.

Some 600 Marines were killed. More than 3,000 were wounded. Another 7,000 suffered from non-battle related injuries, primarily frostbite. Nearly 200 men were missing in action (MIA)—they did not return from the military operation yet it was not known if they were killed or captured.

The losses for the NKA and the Chinese were even higher, with an estimated 12,000 killed and some 10,000 wounded. The Chinese plan to wipe out the US Marine forces had failed— despite vastly outnumbering them.

The courageous fighting during this two-week-long battle produced 17 Medals of Honor and 78 Service Cross Medals.

The battle changed the UN force's expectation of victory. Capturing North Korea and turning Korea into one unified country now seemed like an impossible goal.

Meanwhile, the Communists renewed their mission to defeat the US and UN armed forces.

The Korean War was about to heat up even further.

WHO FOUGHT

Ensign Jesse Leroy Brown

JESSE LEROY BROWN was born in Hattiesburg, Mississippi, into a sharecropper family. He was an athlete who excelled at math and dreamed of becoming a pilot.

In 1944, he left Mississippi to attend Ohio State University. To earn money for his education, Brown worked a midnight shift loading boxcars for the Pennsylvania Railroad. Brown also joined the Naval Reserve. Although a new naval aviation program was recruiting pilots, Brown was told he would never make the cut.

Instead of giving up, he persisted through five hours of written tests. That was followed by oral tests and a thorough physical exam. Brown passed with flying colors.

At age twenty-two, Brown became the first African American to complete Navy flight training. He later became a section leader and flew a Vought F4U-4 Corsair.

In October 1950, he was sent to Korea to assist UN forces.

On December 4, 1950, Brown was flying toward the forces at the Chosin Reservoir when he radioed his squadron: "I think I may have been hit. I've lost my oil pressure."

Brown was forced to crash-land his Corsair in the snow on the side of a mountain. His squadron commander, Lieutenant Thomas Hudner, Jr., circled the crash site in his own Corsair. But Brown didn't emerge from his cockpit. Hudner crash-landed next to Brown's wrecked Corsair, risking his own life to help Brown.

Hudner found Brown trapped by a damaged instrument panel. He was bleeding heavily. Hudner radioed for help, but Hudner could do nothing to get Brown out of the Corsair's wreckage. Just before dying, Brown asked Hudner to tell his wife, Daisy, how much he loved her.

As night fell, and the danger of capture became more likely,

Hudner and a helicopter pilot who arrived later realized there was no way to safely recover Brown's body. Brown's shipmates decided to honor him with a warrior's funeral. On December 7, 1950, seven aircraft piloted by Brown's friends flew several low passes over his downed Corsair. They were close enough to see white snow on Brown's black hair. Reciting the Lord's Prayer, the pilots dropped napalm—a flammable gel—on Brown's plane, cremating his body before the enemy could capture it.

Ensign Jesse Brown was posthumously awarded the Distinguished Flying Cross, the Air Medal, and the Purple Heart. Hudner also received the Medal of Honor for "exceptionally valiant action and selfless devotion to a shipmate."

INTERNET

Here's a website with some interesting photos of the Battle of Chosin Reservoir: kids.kiddle.co/Battle_of_Chosin_Reservoir

More images and some other information from the Public Broadcasting Service: pbs.org/wgbh/americanexperience/features/chosin-battle

Read more about Lt. Thomas Hudner's heroic attempted to rescue Ensign Jesse Brown: navytimes.com/veterans/2017/11/14/how-this-fighter-pilot-crash-landed-his-plane-to-rescue-the-navys-first-african-american-aviator

MOVIES

Retreat, Hell! (1952)

LEWIS "CHESTY" PULLER

June 26, 1898 – October 11, 1971

YOU'VE READ ABOUT the amazing heroism of the Marines who fought in the Battle of Chosin Reservoir. Each of those heroes admired another heroic Marine. His name was Lewis "Chesty" Puller.

Some legends say he earned his famous nickname "Chesty" because of his big, thrust-out chest. Another story about Chesty

Puller insisted his chest was shot away in battle and reconstructed from steel. Still others said "Chesty" was an old Marine expression that described someone with a confident swagger.

Whatever the origin of his nickname, Chesty Puller became the most decorated Marine in US history. Among his many honors were five Navy Crosses (the only Marine ever to win that many) and an Army Distinguished Service Cross.

But despite all that military "metal," Chesty never considered himself better than his men. In fact, he focused on lifting up the Marines who served under him.

"That man made us feel invincible," said one machine gunner who served under Puller.

Lewis Puller was born in West Point, Virginia. His father worked as a grocer but passed away when Lewis was ten years old. Growing up, Lewis listened to old veterans tell stories of fighting in the US Civil War. He went on to attend the Virginia Military Institute but left in 1918 before graduation in order to join World War I.

Lewis said he wanted to "go where the guns are!"

Later, inspired by the brave fighting of the 5th Marines at the Battle of Belleau Wood, he enlisted in the United States Marine Corps. During the years between WWI and WWII, Puller rose through the ranks fighting in jungle battles from Haiti to Nicaragua.

In 1941, with WWII in full force, Puller was given command of 1st Battalion, 7th Marines of the 1st Marine Division. He and his brave men fought in many fierce WWII Pacific battles, including Guadalcanal.

In the Battle of Guadalcanal, then-Lt. Col. Puller and his men landed on the island in September 1942. Puller was the proud commander of a battalion which was almost all young and all inexperienced fighters. In fact, Puller was the only man in the unit with combat experience. Amid the deadly and relentless fighting in the tropical jungle full of enemy Japanese soldiers, Puller kept his men calm. During their first night, the Marines came under heavy naval bombardment. Many of the men, due to no experience in battle, failed to dig foxholes. Puller yelled for his men to keep their heads down and stay under cover.

When the bombardment finally ended, Puller strolled among the shaken men, calmly smoking a pipe and offering practical advice as though they'd all been through a completely ordinary night—when the Battle of Guadalcanal would turn into one of WWII's most deadly fights.

Chesty Puller

In another battle, three of Puller's companies were surrounded by a much larger Japanese force. The companies were completely cut off from help—no reinforcements could be notified. Puller fought his way through enemy territory to the island's shoreline then signaled a US Navy destroyer out on the water to direct fire support. That firepower allowed landing craft to deliver help that could rescue the isolated Marines. Without Chesty's bravery under fire, his men would have been wiped out or taken prisoner by the Japanese.

In still another battle on Guadalcanal, in which Puller earned a third Navy Cross, his Marines were defending Henderson Airfield when a horrific firefight broke out. The Japanese killed more than seventy Americans over a three-hour period. However, the outnumbered Marines who were under Puller's command also managed to inflict more than 1,400 casualties on the enemy—and still hold control of the airfield.

Puller rose to executive officer of the 7th Marine Regiment. He won his fourth Navy Cross for overall performance of duty. Several battalion commanders were under heavy machine-gun and mortar fire when Puller expertly reorganized the battalion—thinking with utter clarity amid the noise, blood, and chaos of battle—then led a successful attack against heavily fortified Japanese defensive positions. The man seemed fearless.

In early 1944, as Allied forces continued to fight its unforgiving Japanese enemy, Puller was promoted to colonel and named commander of the 1st Marine Regiment. Later that year, Puller led the 1st Marine Regiment into a wicked and extended battle on the island of Peleliu. This battle would become one of the bloodiest in Marine Corps history. Under Puller's command, the 1st Marine Regiment lost more than half of their

3,000-man force against the well-entrenched Japanese. Despite the heavy losses, Puller launched continuous frontal assaults. Puller's regiment didn't leave the battlefield until the corps commander ordered Puller to do so.

Puller stuck with his men. And earned their profound respect.

Once he saw a second lieutenant ordering an enlisted man to salute him 100 times simply because the man had missed one salute. Puller told the officer, "You were absolutely correct in making him salute you 100 times, Lieutenant, but you know that an officer must return every salute he receives. Now return them all, and I will keep count."

Chesty Puller quotes became legendary. Here are a few:

"Don't forget that you're First Marines! Not all the Communists in hell can overrun you!"

"Hit hard, hit fast, hit often."

When Puller first saw a flamethrower demonstration, he asked, "Where do you put the bayonet?" He was always ready to get close to the enemy!

But his most famous quote probably came during the Battle of Chosin Reservoir. As you know, the freezing cold Marines were isolated in the mountains and outnumbered by a much larger Chinese army. Their escape seemed impossible. Standing in the icy cold, Puller looked at his men and said, "We've been looking for the enemy for some time now. We've finally found him. We're surrounded. That simplifies things."

Puller's bold attitude was perfectly suited for the transformation of the Marine Corps. The force had begun as a mostly reserve element during the Civil War but continued to rise to towering heights of bravery during World War I, World War II, Korea, and today's modern warfare encounters. At the outbreak of the Korean War, Puller, as commander of the First Marine Regiment, was part of the landings at Inchon—where he earned the Silver Star Medal. His unwavering courage during the Battle of Chosin Reservoir helped turn a very likely disaster into a salvageable retreat that saved thousands of lives.

In May 1951, Puller returned to the United States and took command of the 3rd Marine Division at Camp Pendleton, California. Four years later, he suffered a stroke and was retired by the Marine Corps.

On November 1, 1955, he received a promotion to lieutenant general. He passed away in 1971.

But to this day, the mascot of the Marine Corps is always named "Chesty Pullerton." And the mascot is always a purebred English bulldog.

BOOKS

Chesty Puller's Rules of Success by Bill Davis, Col., USMC (ret) explores 20 of Puller's "self-imposed principles of action."

Chesty: The Story of Lieutenant General Lewis B. Puller, USMC by Col. Jon T. Hoffman, USMCR

MOVIES

John Ford directed the 1976 documentary *Chesty: A Tribute to a Legend.*

The HBO miniseries *The Pacific* (2010). Puller is played by American actor William Sadler.

THE BATTLE OF SAMDONG-NI

February 8, 1952

Douglas AD Skyraiders, McDonnell F2H Banshees, Grumman F9F Panthers, and Vought F4U-5N Corsairs aboard the USS *Essex*, Korea, January 1952.

IF YOU'VE EVER looked out from the top of a very tall building, you'll notice how small everything seems on the ground.

Imagine you're even higher than that—you're flying a plane over a vast ocean. The only place for you to land the plane is a small dot bobbing on the wind-whipped water.

That dot is an aircraft carrier.

The Battle of Samdong-ni centers around one aircraft carrier, the USS *Valley Forge*. Brave pilots flew their planes off the carrier's deck, attacked remote bridges inside enemy territory, then flew back to the carrier—awaiting their next mission.

To these pilots, such extreme danger was all in a day's work.

But this battle also showcases some other unsung military heroes—rescue pilots and aircraft crews—many of whom flew into heavy enemy fire without any weapons except for their personal protection rifles and sidearms.

In early February 1952, the USS *Valley Forge* was anchored off the North Korean coast. The carrier's pilots were given the mission to take out the bridges of Samdong-ni.

Skyraider, Corsair, and HO3S "Horse" helicopter (at front) on the flight deck of USS *Valley Forge* off Korea, 1951.

This mission posed extreme risk to the pilots. Reconnaissance reports showed more than 50 anti-aircraft guns were defending these bridges. The enemy also had radar detection which helped the guns hit incoming aircraft more accurately.

If these pilots wanted to destroy the bridges of Samdong-ni, they would have to fly straight into heavy flak—enemy fire—and pray to God they survived.

Even more challenging, the position of these bridges would force a "glide-bombing" attack. That's very different from a dive-bombing strike.

When dive-bombing, a pilot flies straight down toward his target, releases the bomb, and flies straight back up into the sky. In a glide-bombing attack, the pilot flies his plane toward the target on a horizontal level, giving the enemy much more time to hit a much larger target. And if an enemy is equipped with radar, he can prepare for the attack ahead of arrival. Radar detection will even tell him which direction the plane is attacking from and its approximate flying altitude.

These glide-bombing planes needed to attack the Samdong-ni bridges from an altitude of 1,000 feet and remain over their targets for fifteen to thirty seconds.

US Navy Douglas AD Skyraiders dropping their bombs during the Korean War.

If you don't think that's a long time to stay over a target, close your eyes and count out loud: "One-one-thousand, two-one-thousand, three-one-thousand..." all the way up to "thirty-one-thousand." Imagine enemy fire exploding all around you, but you cannot change course or fly away.

These pilots exhibited extreme courage.

The planes in this mission were Panther F9F fighter jets and propeller-driven AD Skyraiders. You may wonder why anyone would use propeller planes for such a dangerous mission when they had fighter jets. Well, the Skyraiders could carry 2,000-pound bombs powerful enough to destroy a bridge. The jets would attack the radar-controlled anti-aircraft guns and give some protection to the slower, propeller-driven Skyraiders.

Before we get to the action of this mission, you need to know about a second twist to the situation.

Two months before this raid, Lieutenant Harry Ettinger was flying one of those AD Skyraiders south of the town of Wonsan when he was suddenly shot down by enemy anti-aircraft fire. Before his plane crashed, Ettinger bailed out. He was presumed dead.

But just as this mission was about to start, Army Intelligence learned that Ettinger was alive. Some anti-Communist North Koreans had saved the pilot's life.

The mission now shifted. In addition to bombing the bridges, the goal was to rescue Ettinger, too.

Helicopters are the best aircraft for rescue missions. They are agile, can land and take off from rough terrain, and they can hover over a pinpointed location then descend in a straight line. Also, in this case, the anti-Communists who saved Ettinger were asking for a radio. Dropping a radio by airplane would require a parachute that would make the radio vulnerable to getting lost in the wind or intercepted by enemy forces.

Sikorksy HRS-1 helicopter hovers near the ground in Korea while Marines hook a cargo net loaded with 1,000 pounds of supplies.

The helicopter pilot chosen for this rescue mission was Chief Petty Officer Duane Thorin. He would fly a Sikorsky HO3S-1 from the USS *Valley Forge* accompanied by three Corsairs and three Skyraiders that would provide cover fire. This Sikorsky was a small helicopter that could only carry a limited amount of weight. Even before takeoff, Thorin was concerned about that. Onboard this mission would be an intelligence officer, a medical kit, and the radio. Thorin would also be flying the maximum range for the helicopter so he had only had enough fuel to reach Ettinger and immediately return to the ship. Thorin was adamant that the rescue would only succeed if the radio and medical equipment were dropped off *before* Ettinger boarded the chopper and added more weight.

Otherwise, the chopper would struggle at liftoff.

There was no room for error in this mission.

Thorin piloted his helicopter toward the pickup site, flying over anti-aircraft positions, small arms fire, and mountainous terrain full of turbulent winds. He made it.

But suddenly the mission collapsed.

Ettinger—thrilled to see rescue arriving—raced from his hiding place and jumped into the helicopter before the radio and other gear could be unloaded. His extra weight tilted the rising chopper off-balance. The rotor blades sliced into the ground, flinging chunks of dirt and debris into the air. Within mere moments, the helicopter crashed.

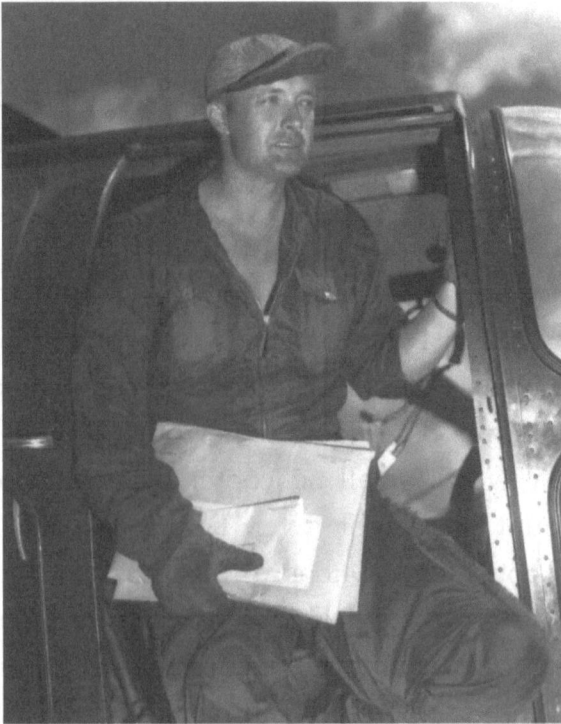

Chief Petty Officer (Aviation Pilot) Duane Thorin

Meanwhile, as the chopper went down, the mission to bomb the three bridges at Samdong-ni was about to launch.

The men in the helicopter had escaped before the crash, but now they were in enemy territory and under enemy fire from the North Korean troops. The Corsairs that accompanied the chopper blasted the enemy with heavy machine-gun fire. The North Koreans retaliated with their own firepower, hitting two of the six planes—one Corsair went down over the ocean, and one Skyraider made it back to a UN-friendly air base.

The remaining four planes circled the helicopter's crash site. The pilots continued to provide cover fire for the chopper crew, trying to keep the enemy at bay, but eventually, the planes ran out of ammunition. And the fuel tanks were almost empty.

Back at the bridge attack, the propeller-driven Skyraiders glided down to drop their bombs. These pilots thought the Panther jets had wiped out the North Korean gun emplacements. But suddenly the heavy anti-aircraft fire opened up. Flak burst around the planes. Despite the enemy fire, the pilots kept flying in and out, holding a horizontal position for fifteen to thirty seconds, and dropping their bombs from a thousand feet altitude.

Their actions were the definition of courage under fire.

The Skyraiders managed to take out two of the three bridges. But with one bridge remaining, they had to fly in again—and they succeeded in blowing up that bridge, too. Unfortunately, one pilot's plane was hit.

The three destroyed bridges of Samdong-ni.

Another chopper swooped in to pick up Thorin and Ettinger. But heavy enemy fire prevented the chopper from landing—in fact, that chopper got hit so badly it barely made it back to a ship near the *Valley Forge* for an emergency landing.

The situation was dire. Two helicopters were downed. One pilot's plane was damaged in the mission to destroy the bridges. And that plane's pilot, Ensign Martin Broomhead, was also wounded by shrapnel—pieces of metal embedded in his skin. After his plane was hit, Broomhead tried to bail out, but his plane was losing altitude so quickly his parachute couldn't have opened in enough time to be effective. As his fellow pilot followed his failing aircraft, Broomhead spotted a snow-covered mountaintop.

He crash-landed.

The powerful impact fractured Broomhead's back and broke both of his ankles. Somehow he still managed to climb from the cockpit, lie down near the plane's wing, and wave at this fellow pilots to signal, *I'm alive.*

The Corsairs providing cover fire were running low on fuel. The two rescue choppers were downed. And the nearest working helicopter was about 100 miles away. For now, Broomhead was on his own.

But when the pilots arrived back at the aircraft carrier and heard about these situation, every pilot onboard offered to fly cover to help rescue the downed choppers and save Broomhead. Tragically, by the time the pilots returned to these crash sites, the men were gone, taken prisoners by the North Koreans. These POWs were not released from captivity until the end of the Korean War.

WHO FOUGHT

Duane Thorin was born in Nebraska to a Swedish immigrant family. Raised on a ranch, he grew up a working cowboy. The discipline stayed with him.

"ON THE *ROCHESTER*, years later, [Duane Thorin] used to entertain we youngsters with rope tricks," said Thorin's shipmate Earl Lanning. "Then, he would pick up a guitar and sing cowboy songs. The warmth of this man was beyond comprehension. We were all in awe of him. He never had to give an order. We all thought he could have flown off the fantail without the aid of a helicopter."

In 1939, Thorin enlisted in the Navy as an apprentice seaman. He was among the few enlisted men for whom pilot training was available (most pilots are officers). As test pilot, Thorin flew every carrier aircraft in service at that time. During the Korean War, he was known officially as Designated Helicopter Pilot #216. Thorin completed hundreds of evacuations and rescues from enemy territory in North Korea.

In February 1952, Thorin was making the rescue attempt you just read about when he was captured by the enemy. He escaped from the POW (prisoner of war) camp in July of that same year but was recaptured. At the end of the war, in September 1953, Thorin was returned to the United States.

After the Korean War, Thorin devoted time to studying the Communists' methods with POWs. Thorin's information helped develop a training program to help prisoners survive, escape, or evade enemy capture.

Thorin's actions during the Korean War earned him the Silver Star. The citation reads:

"The outstanding courage, initiative, and gallantry he displayed in attempting to save the life of another at such great risk to himself, contributed immensely toward maintaining the high morale of airmen participating in action against the enemy, and reflects the highest credit upon Chief Aviation Structural Mechanic Thorin and the United States Naval Service."

BOOKS

The Bridges at Toko-Ri by James Michener—a fictional account based
the Battle of Samdong-ni

INTENET

Here is the film used to train F4U Corsair pilots during WWII:
youtube.com/watch?v=R4aPk4fledU

MOVIES

The Bridges of Toko-Ri (1954): This is a fictionalized version of battle,
so not every detail is accurate. But the movie is a classic military
aviation film.

KEY FIGURES IN THE KOREAN WAR

"THE MILITARY DON'T start wars," said one US general. "Politicians start wars."

That statement by US General William Westmoreland proved true with the Korean War.

While soldiers died all across the Korean Peninsula, politicians stayed far away from the battlefields and argued over ideologies—one side believing in open democracies and the other side wanting oppressive Communist regimes to control people's lives. These clashing political ideas triggered the deadly battles.

Since politicians play such an important role in wars, you ought to know about some key political figures in the Korean War.

JOSEF STALIN

JOSEF STALIN WAS the dictator of the country known as Russia before a violent revolution. After that, Russia joined the Union of Soviet Socialist Republics, a Communist-controlled band of countries in eastern Europe known as the USSR.

During Stalin's twenty-four years as dictator, he was responsible for the deaths of about 40 million people.

It's hard to imagine how one person could be responsible for so many deaths. But Stalin's orders had devastating effects on the entire country. For instance, he sold his people's food supplies to raise money for his own projects. Without food, millions of people starved to death. Adding to that number were Stalin's mass executions as he constantly killed off his "enemies."

Between 1936 and 1938, Stalin instigated "The Great Purge" or "The Great Terror." His goal was to rid the Communist Party of anyone who disagreed with him. To achieve that goal, Stalin had his police execute all "enemies of the state" or send those people into forced labor camps where life was so difficult that about one million of them died there. Stalin also imprisoned or killed many of his Red Army leaders, including generals, field commanders, and naval admirals.

Stalin was a monster. And his people lived in terror of him.

At the start of the Korean War, Stalin wasn't supportive of this invasion of the peninsula. He was concerned that the invasion might launch World War III, and he did not want to fight the powerful United States. Unfortunately, during the late 1940s, an American political leader gave a speech that made it sound like America would not use its military to defend South Korea. After that speech, the invasion didn't appear as dangerous to Stalin. After all, South Korea's military didn't stand a chance against more powerful enemies, such as the USSR and China. If the Communists invaded South Korea, it now looked like they could take control of the entire peninsula.

With Stalin's agreement, North Korea's dictator Kim Il-Sung invaded South Korea on June 25, 1950.

Stalin soon realized his mistake. The invasion launched the Korean War—and the US defended South Korea.

MAO TSE-TUNG

MAO TSE-TUNG (PRONOUNCED "mow say tong") was dictator of Communist China—and just as murderous as Stalin.

China also had suffered through a political revolution during the early 1900s. The revolution destroyed China's monarchy—its kings and nobles—and created a government power ruled by Communists. As a young man searching for a job, Mao was an early member of the Chinese Communist Party.

Rising through the ranks, Mao later led a peasant army revolt whose ideas aligned with Stalin's USSR.

But Mao wasn't Stalin's puppet. When China fell into a bloody civil war—a war fought between countrymen rather than an outside enemy—Mao cleverly established the People's Republic of China. He won converts to his Communist leadership by doing good things. For instance, he expanded public schooling. He offered poor people free healthcare. He promoted equal rights for women.

However, all those good things came at a steep price. Mao imprisoned and killed his enemies just as Stalin did. Also like Stalin, Mao was responsible for causing mass starvation. Millions of people died under his Communist rule.

In 1950, Mao sent Stalin a secret message. He wanted the Communists to invade South Korea—right away. Otherwise, democracy would spread and the Communists would lose power.

"If we allow the United States to occupy all of Korea," Mao wrote in his secret message, "Korean revolutionary power [the Communists] will suffer a fundamental defeat, and the Americans will run more rampant, and have negative effects for the entire Far East."

Mao and Stalin joined forces to invade South Korea and control the entire Korean Peninsula.

KIM IL-SUNG

JAPAN OCCUPIED THE Korean Peninsula for decades, until the end of World War II. But during that occupation, guerrilla warriors were fighting the Japanese army for control of the area.

Guerrilla (pronounced "guh-rill-ah") warriors are unofficial military forces. Often operating as rebels, armed civilians, or militia, guerrilla armies launch small-scale actions such as ambushes and sabotage as part of an overall political and military strategy. One of the fiercest guerrillas fighting against Japan's occupation was headed by Kim Il-Sung.

When Japan lost the Korean Peninsula, Kim was named leader of the newly formed North Korea.

As a young man, Kim had joined the Chinese Communist party. Like Mao, he proved himself to be a ruthless military leader and rose through the political ranks. Kim was also like Mao in that he made himself popular among ordinary people by appearing to fix some of their problems. He took control of land ownership and redistributed money, making the poor richer (and the rich poorer). Kim also resorted to extreme violence and cruelty to maintain his hold on the country. His policies also caused widespread famine—mass starvation—among his people.

Although North and South Korea fought border skirmishes for several years along the 38th parallel, Kim wanted a full-scale invasion that would give him control of the entire peninsula. He asked for Stalin's help for many years and finally got his wish in 1950.

Remember the saying: "Be careful what you wish for."

The Korean War devastated North Korea. The country never fully recovered from it. But Kim continued to rule the country until his death in 1994.

After Kim's death, his oldest son, Kim Jong-il, took control of the country.

HARRY S. TRUMAN

IN 1945, HARRY S. Truman became vice president of the United States.

He served only 82 days.

On April 12, 1945, President Franklin Delano Roosevelt (FDR) suffered a stroke and passed away. That same day, Vice President Truman was sworn in as president of the United States.

Several weeks later, Europe surrendered to the Allied forces fighting World War II.

But Japan refused to surrender.

One of Truman's first orders as Commander in Chief was to send repeated warnings to Japan: Surrender or suffer terrible consequences.

Japan continued to refuse.

In August 1945, Truman ordered two atomic bombs to be dropped on Japan. The explosions burned entire cities to the ground. Soon after, Japan surrendered.

WWII was officially over.

But just a few years later, North Korea invaded South Korea. And Truman, a passionate anti-Communist, faced another difficult challenge as America's Commander in Chief. After the invasion, Truman reportedly said, "By God, I'm going to let them [North Korea] have it!"

The US Constitution—which is the law of the land—states that presidents must ask Congress for a "declaration of war" before entering any major military engagement. However, for the first time in US history, the president did not.

Truman went to the United Nations. The UN then sent US military forces and troops from many other countries to South Korea. These troops were under the leadership of General Douglas MacArthur.

As you now know, these early troops were not prepared to fight overwhelming numbers of Communist forces armed with Russian weaponry. Instead of pushing the Communists back across the 38th parallel, the Communists ran the UN forces to the southern tip of the Korean Peninsula. MacArthur then countered with his brilliant Inchon offensive, which did drive the Communists north of the 38th parallel and liberated South Korea. But it didn't hold.

Chinese troops poured into North Korea.

MacArthur desperately wanted the US to go to war with China. MacArthur also wanted to use nuclear weapons to defeat the Communists. But Truman disagreed. He removed

MacArthur from command, and the Korean War bled on.

The war continued until Truman left office. In 1953, Dwight D. Eisenhower became the president. Eisenhower was an extraordinary military leader who had led the Allied forces to victory during WWII. As soon as he took over as president, Eisenhower pressed for an end to the Korean War. In 1953, he pulled US troops out of Korea. The border between North and South Korea returned to the 38th parallel.

Which goes to show that politicians can start wars, and they can stop them, too.

BATTLE OF THE TWIN TUNNELS

January 29 – February 1, 1951

Infantryman from US Army's 3rd Division mans a 75mm recoilless rifle covering the Tenth Corps in Korea.

YOU MAY HAVE heard the expression "Be careful what you wish for."

This next battle—which is actually two battles—fits that saying.

But first, let's recount what's happened so far. The North Korean invasion succeeded in breaching the 38th parallel. When the first US forces arrived, they were driven to the edge of the Korean Peninsula. But in September 1950, the Inchon landings succeeded and saved the city of Seoul. However, soon after that success, the Communist Chinese joined forces with the NKA and launched a massive counterattack, pushing the UN forces all the way back to the Pusan perimeter—making it seem as if the Inchon landings had gained nothing.

As you can see, this war kept shifting back and forth, over and over.

Now we jump forward to January 1951. US Lieutenant General Matthew Ridgway had devised a plan to hold the defensive line at the Pusan perimeter and also launch another offensive attack against the Chinese Communists and the NKA.

Ridgway's plan was code-named Operation Thunderbolt.

The plan contained two parts: first, a reconnaissance mission to gather information, and second, the Battle of Chipyong-ni to wipe out the enemy.

For the recon mission, Ridgway sent out a patrol unit to locate the Chinese 42nd Army. Ridgway needed to know where this enemy force was because it was so large and so powerful it could determine the success or failure of Operation Thunderbolt.

On the morning of January 29, Lieutenant James Mitchell headed out with forty-four men from the US Army's 23rd Infantry Regiment. These soldiers planned to meet up with

Lieutenant Harold Mueller and his fourteen men from the 24th Infantry Division. Together, their mission was to get recon on the Twin Tunnels—named because one railroad tunnel led directly into the next tunnel.

The recon patrol came well-armed and the men were prepared for action. A fleet of jeeps and three-quarter-ton weapons carriers hauled an assortment of machine guns, recoilless rifles, bazookas, and mortars.

Machine gunners surveil Communist Chinese movements in Korea.

Two jeeps led the front of the recon unit, traveling toward the Twin Tunnels. Nearby was a large hill, numbered 453, along with a stream of water and another rise, Hill 333. At first, everything seemed quiet enough for the vehicles to continue leading the patrol in a forward position, then waiting for the rest of the jeeps to follow.

But one soldier in a forward jeep suddenly spotted movement on a hill.

The enemy!

Mueller's and Mitchell's men opened fire. The Communist forces fired back. Mortar rounds exploded around the jeeps. Dirt burst around the men as they raced for the nearest cover.

The best defensive position is always the high ground. Troops can see the enemy approaching and fire down on them.

Mueller ordered his men up the steep snowy mountainside of Hill 333. So did Mitchell. But enemy soldiers were also racing for the rise. This scramble uphill turned into a fierce footrace between fifty-one US soldiers and hundreds of Communist soldiers who were right on their heels. Mueller's men finally reached a ridge about 400 feet above the road. But the Communists outflanked them by taking even higher ground. Not only could the enemy now fire down on Mueller's men, but the ground was frozen. The men couldn't dig in for cover.

Mitchell's men spread out to hold this one part of the ridge. Their defensive weapons consisted of one .30-caliber machine gun, one bazooka, and eight Browning Automatic Rifles.

As the men set up their defensive perimeter, the enemy opened fire with mortars. Then they attacked.

The US soldiers fired back and halted the enemy in their tracks.

But a Communist machine-gun crew climbed to a position directly above the Americans and rained fire down on them. US Sergeant Everett Lee crawled from his cover, spotted the enemy, and opened fire. Everett—an expert marksman—killed two soldiers in the machine-gun nest. Inspired by Everett's example, other GIs (the nickname for General Infantry soldiers) also opened fire. The machine gun fell silent.

But this enemy assault was just the beginning.

By that afternoon, seven Americans were killed or wounded.

Fortunately, a small Army scout plane was flying overhead and reported the situation to Colonel Paul Freeman, commander of the 23rd Infantry Regiment. Freeman realized these men were vastly outnumbered. He ordered an air strike.

On the mountain, the GIs heard the ear-rattling thunder of jet engines. Moments later, rockets exploded among the enemy forces. Then the jets opened up with strafing machine gun fire.

Air Force F-80 Shooting Star releases its bomb load (below left wing).

Soon, four more jets approached. Red flames exploded among the Communists. The jets were dropping napalm—the explosive gel that sticks to any surface before burning it to ashes.

But three-and-a-half hours after this hasty scramble into a defensive position began, one-third of the American force was casualties. And their ammunition, food, and water were running low.

Another plane was sent to drop ammo to the men, along with a hopeful message: A relief column of 167 men was heading their way, led by Captain Stanley Tyrrell.

Corsair turns to survey the damage inflicted (upper center).

The situation for the Americans seemed to be improving.

But during the night, the Chinese launched mortar and small arms fire, followed by shrieking whistles, blaring bugles, and men screaming war cries. However, amid that attack, three men of Mitchell's patrol slipped through enemy lines. Following railroad tracks, they headed toward the town of Sinchon, about a mile from the Twin Tunnels. Amazingly, they ran into Captain Tyrrell's relief column!

Good news, right?

Almost.

These three men—who had witnessed the massive Chinese forces running up Hill 333—told Tyrrell's relief column that it was unlikely anyone else was alive on the hill. Also, the Communist attack had just stopped. The deafening silence told them nobody else survived the attack.

In the dark, Captain Tyrrell set up a new defensive perimeter. About a half-hour later, one of Tyrrell's platoon leaders received a radio message. Another man from the hill just stumbled into their lines. He said they'd taken heavy losses but other men were still alive on the hill.

To help retrieve the GIs on the hill, Tyrrell sent out a reinforced platoon.

Back on the hilltop, the Communists launched their fifth attack. The Americans held—only to face two more attacks. But after seven assaults, the Americans' ammunition was nearly gone. Casualties ran as high as 75 percent of the men. And most of the command had been wiped out.

The situation was desperate.

And now the men on the hilltop heard footsteps, crunching through the ice and snow. Exhausted, depleted of ammunition, the men prepared for their eighth attack. Surely this one would mark their end on earth.

Then a voice cried out in the freezing darkness: "Don't shoot! It's GIs!"

The relief column!

Fifty-nine men of Tyrrell's patrol arrived. Each one was shocked by the condition of the men on the hill. Only about ten men had not been seriously wounded. And the fight wasn't over.

Now the battle moved to Chipyong-ni—which is the subject of our next chapter.

BATTLE OF CHIPYONG-NI

February 13–15, 1951

Army artist's depiction of the of Battle of Chipyong-ni.

REMEMBER THAT SAYING, "Be careful what you wish for?"

Lieutenant General Ridgway wished he knew the location of the Chinese 42nd Army, and now he got his wish. The deadly and powerful enemy force was advancing right toward the Twin Tunnels and the village of Chipyong-ni. For Ridgway's attack plan to succeed, he needed to stop them.

Ridgway ordered General Edward Almond of the 4th Corps to set up a defensive line at the Twin Tunnels. He also put Freeman—who, in our last chapter, ordered the air strike to save those lives on the hill—in charge of four battalions of the 23rd Infantry Regiment along with a French battalion. The French force was led by Lieutenant Colonel Ralph Montclair (you'll read more about him later).

Freeman stationed his men on both sides of the tunnels, east and west, on Hill 333. The French battalion took the rough terrain of Hill 453. These positions placed the US and French forces across from each other, which allowed for fire support from heavy artillery located about two miles away.

UN forces dug into a hillside in Korea.

But while the UN forces were setting up their positions, the Chinese were watching.

In the early morning darkness of February 1, the Chinese 374th Regiment—part of a 7,000-man division—attacked.

In order to spot the enemy in darkness, Allied forces shot flares into the sky. US and French forces opened fire. Mortars exploded. Machine guns blazed. The hills zinged with red lines from tracer rounds—bullets that light up for better accuracy in the dark.

This overwhelming response took the Chinese by surprise. Their attack paused.

But not for long.

The Chinese forces regrouped, reinforced their lines, and attacked Hill 453 where the French were located.

One clever Chinese tactic was "human waves." The Chinese military had so many soldiers that their leaders never seemed to run out of men. And since those leaders wanted to win at any cost, they didn't really care about the extreme body count of dead soldiers—they simply sent more men to the battlefield.

Nearly smothering the hillside with wave after wave of Chinese soldiers, the UN forces struggled to hold them back. Overcast skies also meant the Allies couldn't bring in air support.

The Chinese broke through the French defensive line.

The fight fell into hand-to-hand combat as men assaulted their enemy with any object nearby—rifle butts, knives, rocks, fists. For three hours the battle raged. French ammunition supplies dwindled to near nothing. The situation looked hopeless.

But rather than surrender, French Colonel Montclair ordered a bayonet charge. Steel blades flashing, the French soldiers lunged toward the Chinese. They drove them off the hill.

Lieutenant Colonel Ralph Montclair

The Chinese weren't finished, either.

Their goal was to split the UN forces and outflank them by coming around the lines from the side. Fortunately, some US tanks launched cannon fire and halted this attack. And when the next Chinese attack came, hitting the US forces with their usual scare tactics of blowing whistles and screaming, the heavy artillery support located two miles away responded by blasting the Chinese troops off the hillsides.

But the Chinese launched yet another assault! Their clear advantage was that they had access to an overwhelming number of soldiers. This time, the Chinese attacked both French and US forces, and they managed to break through one of the US positions on high ground.

Freeman's headquarters were now exposed to enemy fire. The high ground also gave the Chinese an opportunity to fire on other UN positions, including a First Aid station, and to create another opening in the French defensive lines.

Freeman's response? Not surrender.

"Open fire with all weapons!"

The troops unleashed every ounce of firepower, including an M-24 light tank equipped with twin 40mm anti-aircraft guns.

Americans firing on Communists positioned in Korea.

Freeman's goal, he later said, was to "vacuum clean" the enemy's high ground position.

Within about ten minutes, smoke and flames were ripping through the heights of the hilltop. The French forces then surprised them by rushing in.

"Screaming like madmen," Freeman said, "again, they came out flashing their steel bayonets as the Chinese looked in horror."

The Chinese soldiers broke and ran—or died.

And yet this was *still* not the end of the Chinese attacks. Hitting the US forces on the east and west position by the Twin Tunnels, their overwhelming force broke through the defensive perimeter. The fight once again fell into close-quarter combat— bayonets and hand grenades.

The Chinese took the high ground.

It was now 3:00 p.m. The battle had raged for almost twelve hours. Not only were the men exhausted, wounded, hungry, and thirsty—they were cold. In the mountains of Korea, February plunges to temperatures below freezing.

Freeman realized his men were about to be overrun by the Chinese. Refusing to surrender, he regrouped the diminished forces on the east tunnel ridge. Suddenly, the sky cleared. Soon after that, four US Marine F4U Corsair fighter bombers screamed into the air space above them. The planes dropped 500-pound bombs on the Chinese. Corsair pilots followed with rocket and cannon fire. UN artillery fire joined the fight, decimating the enemy.

After that counterattack, ammunition and reinforcements arrived for the US and French forces. The men hunkered down and prepared for the next Chinese attack.

Navy F4U Corsair fighter leaves the deck of a carrier ship operating off the coast of Korea to sortie against Communist-led North Korean forces.

It never came. The enemy had run for their lives.

The Battle of Chipyong-ni was abruptly over.

When Ridgway later recalled this fierce fight in Korea, he offered only the highest praise for his steadfast men.

"Isolated far in advance of the general battle line, completely surrounded in near-zero weather," he said, "they repelled repeated assaults by day and night by vastly superior numbers of Chinese.... I want to say that these American fighting men, with their French comrades-in-arms, measured up in every way to the battle conduct of the finest troops America and France have produced throughout their national existence."

The Battle of Chipyong-ni also boosted the morale of the 8th Army, which had wondered if the Chinese forces were invincible.

On the other side, the Chinese suffered a serious setback. They had hoped to drive the UN forces to the edge of Korea's peninsula. Instead, they were driven back. As Chinese leadership said, "In the conduct of the ... battle, we have underestimated the enemy."

WHO FOUGHT

Lieutenant Ralph Montclair, otherwise known as Raoul Charles Magrin-Vernerey.

RAOUL CHARLES MAGRIN-VERNEREY was born in Hungary on February 7, 1892.

At age fifteen, he tried to volunteer for the French Foreign Legion but was denied entry.

The French Foreign Legion is unlike any other army.

Although Legionnaires, as they're called, are highly trained infantry soldiers, this French army is open to literally anyone who passes its extremely challenging physical and mental training. "Anyone" includes criminals, outlaws, convicts, bloodthirsty mercenaries (men who fight battles solely for money), or someone who isn't even a French citizen.

The Legion is also the only section of the French military

that swears allegiance to the Foreign Legion, not to the country of France. Any Legionnaire—for example, an outlaw from Venezuela—who gets wounded in battle while fighting for France can immediately apply to become a French citizen. His citizen status is then known as "French by spilled blood."

After Magrin-Vernerey was turned down by the Legion, he graduated from a special military school and joined the French army's Infantry Regiment. He rose through the ranks. By the end of WWI, he was a captain, having earned eleven military awards and citations. He also was wounded seven times, including a fractured thigh from a bullet, a broken arm from a grenade, and gas burns to his eyes. Most of those wounds would disqualify a man for military service.

Magrin-Vernerey pressed forward.

In 1924, he joined the ranks of the French Foreign Legion.

Battle-scarred and hardened, he was a perfect fit. During WWII, he changed his name to Ralph Montclair—it was easier for Allied soldiers to remember—and proved his mettle again and again on the battlefield.

Once, a young soldier confessed he was afraid of fighting. Montclair replied: "Don't worry. You'll fight. We'll win."

After WWII, Montclair stayed with the Legion. In 1950, at the age of sixty, when most military commanders retire from active duty, Montclair volunteered to command French forces in the Korean War.

But in order to assume command on the front lines, Montclair had to voluntarily demote—lower his military rank—from general to lieutenant colonel. No general wants to move down in rank.

But Montclair didn't hesitate.

Among his many battles in Korea, Montclair was in command of the French battalion that fought alongside the US 23rd Infantry Regiment at the Battle of Chipyong-ni.

BOOKS

Crossroads in Korea: The Historic Siege of Chipyong-ni (Macmillan Battle Books) by T. R Fehrenbach

MOVIES

Documentary *Korea: Chipyong-ni*:
youtube.com/watch?v=_QIgPfH9SG8

Documentary *Korea: Twin Tunnels*:
youtube.com/watch?v=p97eFNwv_Mg

REVEREND EMIL JOSEPH KAPAUN

April 20, 1916 – May 23, 1951

Father Kapaun holds Mass in Korea, using a jeep's hood for an altar, October 7, 1950.

YOU'VE READ ABOUT heroes who fight with blazing guns on the battlefield. These warriors sometimes sacrifice their own lives to save others.

But another kind of warrior fights his enemy without visible weapons. This man faces the gravest dangers on earth armed only with his courageous faith in God. These brave warriors also sacrifice their lives to save others.

Reverend Emil Joseph Kapaun was that kind of warrior.

During the Korean War, Kapaun was a chaplain with the 8th Cavalry Regiment. Military chaplains normally represent one particular religion or faith—such as Christianity—but serve all personnel of every faiths or even no faith at all. Soldiers go to their chaplain when they need spiritual guidance or prayer.

Kapaun was a Roman Catholic priest. During WWII, he served the troops on the battlefield. In 1950, Kapaun was among the first regiments to arrive in Korea.

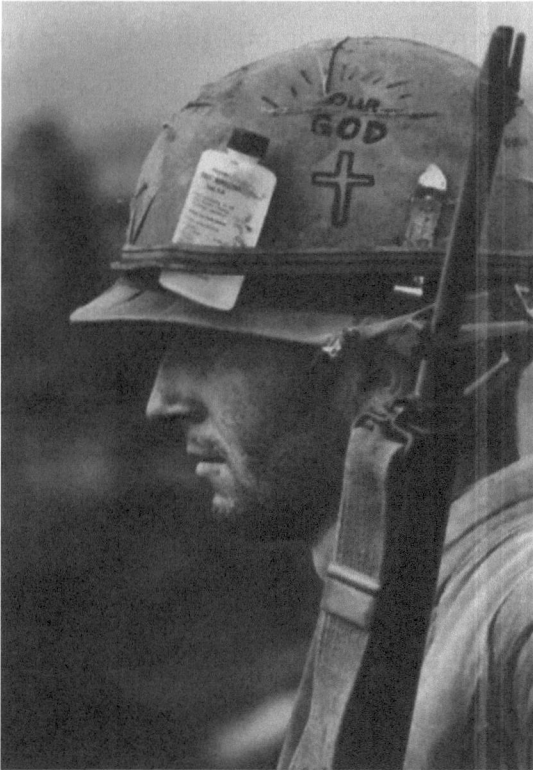

GI's helmet adorned for extra protection.

Kapaun quickly became concerned about the strength of the Communist forces fighting in Korea.

"The Reds are too strong for us," he said. ("Reds" was slang for Communists.)

Kapaun saw the war from the front lines—often under near-constant mortar and machine-gun fire. He was with the men during the Inchon Landing and other crucial battles. Despite barely escaping three enemy attacks that nearly killed him, Kapaun continued to serve the troops without rest. He rescued the wounded and dead from the battlefield, performed baptisms, heard confessions from soldiers, and offered Holy Communion to celebrate Catholic Mass, sometimes using a jeep's hood as an improvised altar.

Kapaun worked so close to the battlefront that enemy fire sometimes destroyed his religious kit for performing Catholic Mass. In his letters back home to friends and family in Kansas, Kapaun said it was the prayers of others that kept him alive during these ferocious attacks.

Despite the challenges, Kapaun kept his sense of humor.

"My pipe got wrecked again as a Red machine gunner sprayed us with lead and we had to hit the ditch," Kapaun wrote in a letter to his brother in October 1950. "It is funny how a fellow can jump so fast into a ditch. This time it did not have water in it. The last time, the ditch had water in it and you can imagine how we looked. We do have a few laughs in spite of the evils of war."

Kapaun and his broken pipe. *The Wichita Diocese*

On November 1, 1950, despite bitter winter weather and increasingly dangerous fighting—plus his own severe lack of sleep—Kapaun still held religious services for the men. On this day, however, Kapaun's religious service was interrupted by an urgent message. Twenty-thousand Chinese soldiers were fast approaching. Kapaun was told to "bug out"—leave immediately—to avoid getting captured or killed.

Military chaplains in the field were often given a jeep with a driver that took the chaplain wherever he was needed. As Kapaun and a battalion of soldiers moved out, they discovered the Chinese had blocked all the main roads. Kapaun's driver attempted several different routes, only to find each one was under constant enemy attack. Further complicating the escape was Kapaun's insistence on stopping the jeep to pick up every wounded man. Kapaun's driver eventually wound up right back at the aid station where they'd started. Kapaun went inside to help with the wounded and dying. The Chinese tightened their perimeter even further.

Wounded warrior carried from the battlefield of Korea.

The situation was steadily worsening.

Inside the aid station, Kapaun and a doctor struggled to save men's lives. Supplies were limited. Enemy fire was increasing. The relentless Chinese military continued to close around the battalion, hurling hand grenades at the aid station and killing some of the wounded men inside. Kapaun and the doctor helped dig trenches to protect the remaining wounded. Then Kapaun had a desperate idea.

Among the wounded men in the aid station was a Chinese officer. Why would UN forces provide care to an enemy solider? Because doctors have a moral obligation to save lives—all lives, no matter which side of the war they came from. However, this code of ethics was mostly followed by the Allied forces, not their enemies. Kapaun persuaded this Chinese officer to ask his fellow Chinese soldiers to stop this assault—the aid station held dying soldiers who had no way of fighting back. The officer made the request, and soon the explosions diminished.

The following day, orders were issued for the UN forces to withdraw.

Reverend Emil Joseph Kapaun

But as Kapaun and the doctor were escaping, Kapaun learned that one wounded soldier was still out on the battle-field. Desperate to find the man, Kapaun and the doctor searched the area—and were captured by the Chinese.

Kapaun, along with other captured men of the 8th Army, was marched nearly 100 miles to a prisoner of war (POW) camp.

The situation continued to worsen.

Life in the Chinese POW camps was absolute torture. Every day, about twenty men died from malnutrition, disease, and extreme cold. Temperatures sometimes plunged to 20 degrees below zero while men shivered in the prison barracks without heat.

Refusing to give in to despair, Kapaun continued to serve

the men. He gave away his own food and dug latrines—holes in the ground used as toilets. Kapaun also stood up to Communist indoctrination.

The Communists continually tried to brainwash the Allied prisoners. In many camps, a Communist radio station—Radio Beijing—would crackle over the camp's loudspeakers. In English, the daily reports would criticize the United States and United Nations, making them the villains who were responsible for the prisoners' misery. These broadcasts also insisted God did not exist and Communism was the answer to a better life.

Meanwhile Kapaun kept hope alive. He continued to lead the men in prayer. He offered religious counsel to Catholics, Protestants, and Jews. Without complaint, he took on the camp's worst jobs, from washing filthy clothes and sick bodies to ensuring the men had safe water to drink by boiling it over open flames.

When the food ran out, Kapaun prayed to Saint Dismas to help him feed the hungry. Dismas was one of the three thieves crucified with Jesus at Calvary. But Dismas realized Jesus was God's son and innocent of any crime. Dismas begged Jesus to forgive him. Jesus replied, "Today you will be with me in paradise."

With his prayers to the reformed thief Saint Dismas, Kapaun began stealing food from the prison guards. After these forages, he would return to the barracks with stolen loot hidden under his weather-beaten overcoat. Kapaun once managed to steal an entire sack of potatoes. Other times his pockets were full of salt—a rare blessing for the starving men who desperately needed sodium to stay alive.

POWs huddle near a stove to keep warm.

But by the winter of 1951, Kapaun's own health was failing. He had given away his food for so long that his body was breaking down from malnutrition and extreme weight loss. One leg developed blood clots. He went blind in one eye. His other eye filled with fluid. And dysentery—a painful bowel disease—gripped him day and night, eventually leading to deadly pneumonia.

In the spring of 1951, the Communists planned to take Kapaun to a hospital from which nobody ever returned. The Chinese Communists denied these patients any medication. Basically, going to this "hospital" was a death sentence.

The camp's prisoners uttered a tearful farewell to their chaplain.

On May 6, 1951, Reverend Emil Joseph Kapaun died. He was thirty-five years old. And a true warrior who sacrificed his life to save others.

In 2013, the United States awarded Kapaun a posthumous

Medal of Honor. The citation reads:

"Chaplain Kapaun's extraordinary heroism and selflessness, above and beyond the call of duty, are in keeping with the highest traditions of military service and reflect great credit upon himself, the 3rd Battalion, 8th Cavalry Regiment, the 1st Cavalry Division, and the United States Army."

INTERNET

From the History channel: youtu.be/jRhpoZ47Zms

MOVIES

The Miracle of Father Kapaun

BATTLE OF IMJIN RIVER

April 22–25, 1951

British forces fire a 25-pound cannon at enemy positions along the Imjin River, 1951.

CHINESE GENERAL PENG Teh-huai needed to redeem himself.

Teh-huai was in charge of that early mission to capture Seoul from UN control. The Chinese general had failed miserably.

But in the spring of 1951, the general devised a new plan.

Teh-huai planned to use the Chinese 63rd Army—a massive force of some 300,000 soldiers—to attack the weakest point in the UN's defensive line. That spot was held by the British 29th Brigade. But the general realized if he tried to move his large body of men and material over major roads, UN air forces would spot them and very likely wipe them out from the sky.

Instead, the Chinese general cleverly used backroads to camouflage his advance. Once he destroyed these particular British soldiers, Teh-huai planned to sweep his way through the rest of the 29th Brigade's rear areas before steadily advancing south. That path would open the road to Seoul where the general could attain his ultimate goal of conquering South Korea's capital city.

The regiment holding the UN's weakest point was known as the Gloucesters—also known as the Glorious Glosters. A highly respected force, the Glosters had fought for centuries in remote places such as India and South Africa. During WWII, the Glosters were part of the Normandy Invasion that helped defeat the German Nazis.

The only reason the Glosters' defensive position in Korea was considered weak was because the men were stretched too thinly over a long distance.

The clash between the Glosters and the Chinese forces resulted in the Battle of Imjin River—a legendary three days and nights that turned into the British Army's bloodiest engagement since WWII.

Men of the Gloucesters advance with Centurion tanks to attack Hill 327, Korea.

Among the Chinese army's most effective tactics was infiltration. Under cover of darkness or natural camouflage such as forests, Chinese soldiers would sneak up as close as possible to the enemy. Once they infiltrated that territory, the soldiers would surround and isolate them, cutting off any help from arriving. Finally, using their vast numerical superiority, the Chinese would send in wave after wave of soldiers.

Before dawn on April 22, the 29th Brigade sent out several patrols that met heavy Chinese resistance. Later, as night fell, UN air reconnaissance also spotted heavy movements of enemy troops—heading right toward the Glosters.

The Glosters were set up on multiple hills. In total there were about 700 men, commanded by Lieutenant Colonel James Carne.

With this new reconnaissance information, Carne told his men to prepare for an attack. He suspected the main area of attack would be at a crossing of the Imjin River. Carne positioned himself there with some of his men.

Sure enough, in the dark of night, Chinese troops waded across the Imjin River. The Glosters fired flares into the night sky, lighting up the sky for a clear view. They opened up artillery fire, wiping out the first wave.

But the enemy soldiers kept coming. And coming. As the Glosters riddled them with bullets from rifles and machine guns, hundreds of dead bodies began floating down the river. At the same time, the outnumbered Glosters redirected artillery and mortar fire across the river to the far bank, where more Chinese soldiers waited to cross the water. This strategy began wiping out the Chinese Communist forces.

"I think the Chinese had expected to just simply carve a way straight through us," Gloster Sam Mercer later recalled.

"They hadn't expected us to resist... as long as we did. Yes, we upset their timetable."

Centurion tank fires on Chinese positions on Hill 327 in preparation for the assault by the Gloucesters.

But the relentless enemy soon found another way to forge the river. Moving further downstream, they attacked the Gloucesters' Company A.

During this six-hour battle, in which Company A was outnumbered six to one, the Chinese managed to capture one bunker on a forward slope. That position gave them a starting point for their next action—attack the crest of the nearest hill.

The only defense left was the Gloucesters' 2nd and 3rd Platoons (platoons are usually about twenty-five men).

The 1st Platoon was sent to reinforce the 2nd and 3rd, but as they moved for the hill's crest, they were attacked by hundreds upon hundreds of Chinese soldiers behind them. The 1st Platoon responded by firing every available weapon and throwing a rolling barrage of hand grenades.

The counterattack worked. The Chinese ran for cover.

The 1st turned their attention to the bunker taken by the Chinese.

Allied soldier using 57mm recoilless rifle to fire on Chinese Communists.

British Lieutenant Philip Curtis was leading the 1st Platoon. He ordered some of his men to give cover fire while he rushed for the bunker. Although seriously wounded in that charge, Curtis made a second attempt and was within just a few yards of the bunker, already throwing a hand grenade, when he was killed.

However, his grenade struck its target. The explosion allowed the 1st to retake the bunker.

But the Gloucesters' Company A was in bad shape, down to

one officer and fifty-three men. The company was ordered to withdraw and head for Hill 235, which they'd nicknamed Gloster Hill. Companies B and D were also taking heavy losses but continued to hold their positions on different hills. Carne decided to withdraw all his men to Gloster Hill. He then repositioned Company B to Hill 314.

The fight never let up. The Chinese general was determined to annihilate these British soldiers.

Imjin River valley and its many steep hills.

Dawn broke. The wounded were withdrawn. Food and ammunition were delivered to the men holding the line.

But as the day stretched on, the Chinese continued to pursue their tactical plan of surrounding and isolating. They now closed off any road that helped the Gloucesters receive reinforcements or allowed for retreat.

General Teh-huai concentrated the next attack on Companies B and C. The Chinese opened fire with mortars and machine guns. But when the first wave of Chinese soldiers moved forward, they took a wrong turn. The British artillery opened fire and inflicted heavy casualties on the enemy.

Companies of the Gloucesters atop Hill 327, awaiting the next attack

The next day Company B was still holding its position. However, during the night, the Chinese had moved onto a ridge that separated Companies B and C. This infiltration was a smart tactical move. By dividing the Gloucesters companies, the Chinese weakened the force.

Next the Chinese concentrated on the Gloucesters' headquarters.

Carne realized his thin line was stretching even thinner. He ordered Companies B and C to move to Gloster Hill. But to reach their destination, these two companies—about 200 men— would have to fight an estimated 3,000–4,000 Chinese warriors.

Meanwhile, a relief force tried to reach Carnie's men twice. Both attempts failed.

The Gloucesters' situation was grim. They were isolated, surrounded, and outnumbered. The Chinese attacks would only keep coming.

Carne was down to 400 men. Food and water were scarce. Ammunition supplies had dwindled, leaving only enough firepower to cover about ten hours of fighting. Batteries were also running low on charges, which threatened communications with the outside. All bad news.

The only good news was that Carne would receive aircraft help the next day.

Carne pulled his troops to the summit of Gloster Hill. This was a smart move. It condensed his perimeter, making it more defensible. The Chinese would need to make a steep climb to attack the summit.

Wrecked bridge over Imjin River.

On April 25, Chinese bugles blared and whistles shrieked. As the Chinese attacked in human waves, the Gloucesters held their positions—again and again and again. The Chinese attacked seven times, the battle raging all day. Carne led defensive counterattacks with rifle fire and hand grenades. Even though they were seriously outnumbered, Carne and the Gloucesters sent the enemy reeling.

Unfortunately, they soon realized radio communications were failing. Artillery support from outside was not available.

The Gloucesters were ordered to withdraw.

Suddenly, the sky overhead fractured with the boom of jet engines.

A squadron of F-80 Shooting Stars blasted the Imjin Valley below Gloster Hill. Pilots dropped bombs and strafed—attacking by aircraft with machine guns—the enemy below.

The Chinese ran for their lives. Carne and his men dispersed and withdrew.

But overwhelming numbers of enemy soldiers remained throughout the valley. During the breakout, Carnie was captured and taken prisoner.

Only one company of Gloucesters—five officers and forty-one men—made it out alive.

In total, fifty-nine Gloucesters were killed. Nearly 600 were taken prisoner along with Carne and remained in POW camps for two more years, until the end of the war. Thirty-four Gloucesters died in the POW camps.

However, in this battle the Gloucesters had inflicted far more damage on the Chinese army. Casualties were estimated at 70,000 men.

Although the Battle of Imjin River was a defeat for UN forces, it proved an important milestone. The outnumbered Gloucesters held the line. And their courageous stand allowed other UN forces time to regroup so they could block any Chinese advances on Seoul. It was now clear to both sides that the UN forces would not be easily defeated.

Weeks later, the Soviets who were backing the NKA forces hinted that maybe it was time to seek a settlement with the UN.

By July, negotiations began to end the war.

Unfortunately, fighting on the ground would continue for two more years.

WHO FOUGHT

Lieutenant Colonel James Carne.

JAMES CARNE WAS born in England in 1906. He was the son of a brewer.

After graduating from Royal Military College, Carne was commissioned in 1925 as a second lieutenant in the Gloucesters. Later, during WWII, he was promoted to major. When the Korean War broke out, Carne was a lieutenant colonel. He was forty-five years old.

Carne led the Gloucesters in several fierce battles and successful counterattacks on the Chinese. The Battle of Imjin River earned him the Victoria Cross, the British highest military honor given for valor "in the presence of the enemy."

Carne's leadership proved crucial to the Gloucesters holding their line during the brutal three-day attack. His Victoria Cross citation reads:

"Throughout [the battle] Colonel Carne moved among the whole battalion under very heavy mortar and machine-gun fire, inspiring the utmost confidence and the will to resist among his troops. On two separate occasions, armed with rifle and grenades, he personally led assault parties which drove back the enemy and saved important situations. His courage, coolness and leadership was felt not only in his own battalion but throughout the whole brigade."

After Carne was captured and taken to a POW camp, the Communists singled him out for the most intense torture. Kept in solitary confinement—no contact with his men—Carne was fed mind-altering drugs and subjected to relentless "re-education" sessions to brainwash him into renouncing freedom and supporting Communism. Carne's two years in captivity were a continuous horror story. Men sometimes broke under that strain. Carne resisted.

After the war ended, the POWs were returned to their countries.

Carne enjoyed another thirty-three years of life in freedom, and passed away at the age of eighty.

BOOKS

Imjin River 1951: Last stand of the 'Glorious Glosters' by Brian Drohan

Glorious Glosters: A Short History of the Gloucestershire Regiment, 1945-70 (Famous Regiments) by Tim Carew

INTERNET

A short video about the battle and the Glorious Glosters:
youtube.com/watch?v=-QqxI9t0Z4I

SERGEANT RECKLESS

Sergeant Reckless under fire during the Korean War.

A FEMALE HORSE was among the bravest fighters in the Korean War. Her name was Sergeant Reckless.

Here is how she earned her military rank.

In October 1952, a Korean stable boy was selling a small chestnut-colored horse. The boy loved the horse, which he'd named Flame, but needed money to help his sister. She had stepped on a land mine, and the explosion destroyed one of her legs. (Wars cause many casualties, even among civilians.)

The horse named Flame was bought by members of the US Marine Corps. They renamed her Reckless and trained her to work as a pack animal for the Recoilless Rifle Platoon of the 5th

Marine Regiment.

This Recoilless unit needed an animal that could carry their heavy 24-pound shells over the rugged and steep mountains of Korea. At times, the horse would need to carry as many as nine shells, which in total was about 200 pounds.

Reckless' main trainer was Gunnery Sergeant Joseph Latham.

Latham and some other Marines taught Reckless many important skills. She needed to know how to avoid getting entangled in barbed wire, how to lie down when they were under fire, and even how to run for a bunker when somebody cried, "Incoming!" Reckless learned it all.

Platoon Gunnery Sergeant Joseph Latham and Sergeant Reckless.

In the beginning, Reckless stayed inside a pasture near the Marines' camp. But her gentle personality charmed the whole platoon and soon she was allowed to roam through camp.

Sometimes she would even enter the Marines' tents and sleep with them. On the coldest nights, she lay next to Latham's tent stove.

Reckless had a huge appetite. The men fed her buttered toast, chocolate, shredded wheat, peanut butter sandwiches, and mashed potatoes. Her favorite food was scrambled eggs, sometimes washed down with her favorite drink, Coca-Cola. But Reckless was also known to eat whatever object was nearby, including poker chips. She once ate her own blanket.

Military camp tents in Korea.

The first time Reckless heard the booming explosion of a recoilless rifle, she jumped straight up in the air—all four feet lifting off the ground. When she landed, she was shaking. After the men calmed her, they fired the gun again. Reckless snorted. By the end of that mission, she was calm—and so hungry she tried to eat someone's helmet liner.

In order to deliver ammo and supplies into battle, Reckless needed to continually learn new delivery routes. The men discovered she only needed a little guidance before she could make the entire trek through foreign territory all by herself.

On the battlefield, Reckless proved a valiant hero.

Sergeant Reckless standing beside a 75mm recoilless rifle.

During one battle, Reckless made fifty-one trips by herself in one day, carrying a total of 386 recoilless rounds—more than 9,000 pounds over rough terrain. She was wounded twice. Shrapnel struck an area above her left eye. Another piece hit her left flank.

Reckless was awarded two Purple Hearts and promoted in rank.

Reckless also packed other items for the platoon. She was particularly useful for stringing telephone wire. She would carry the reels of wire in her packs. The wire would reel out as she walked, laying out more wire than ten men on foot.

For this exemplary service to the Marine Corps in Korea, Reckless was awarded a Marine Corps Good Conduct Medal. She also earned a Presidential Unit Citation with bronze star, the National Defense Service Medal, a Korean Service Medal, the United Nations Korea Medal, a Navy Unit Commendation, and a Republic of Korea Presidential Unit Citation.

Reckless wore these honors on her horse blanket.

Sergeant Reckless getting promoted to Staff Sergeant at Camp Pendleton, 1959. General Randolph M. C. Pate, Commandant of the Marine Corps, tacks the chevrons onto her blanket.

After the Korean War, Reckless was moved to Camp Pendleton, California. She produced three foals: colts Fearless (1957), Dauntless (1959), and Chesty—named for Chesty Puller, who was among the very few Marines ever allowed to ride Reckless.

In 1960, Reckless retired from active service with full military honors at Camp Pendleton. Instead of retirement pay, she was given free quarters and fed like a queen.

"She wasn't a horse," one writer later said. "She was a Marine."

BOOKS

Sgt. Reckless: America's War Horse by Robin Hutton

They Called Her Reckless by Janet Barrett

Reckless: Pride of the Marines by Andrew Geer

Sergeant Reckless: The True Story of the Little Horse Who Became a Hero by Patricia McCormick

INTERNET

Here's a video of Sgt. Reckless with some good images: youtube.com/watch?v=CBY8_-BPIwk

MOVIES

Sgt. Reckless (documentary)

BATTLE OF BLOODY RIDGE

August 18 – September 5, 1951

DO YOU KNOW what "persistence" means?

Persistence is continuing to do something no matter how difficult.

In this next battle, UN forces showed tremendous persistence. They never gave up, despite suffering terrible losses.

In fact, the Battle of Bloody Ridge became the bloodiest fight of the Korean War.

By July 1951, peace talks had taken place between countries, but the discussions had fallen into a stalemate. Neither side wanted more trouble, but neither side agreed to stop attacking the other.

With the threat of attack continuing, both sides rushed to reposition their lines into the best defensive positions. That strategy required taking the high ground.

As you know, Korea is very mountainous. Much of the high ground was taken by the Communist forces who were firing down on UN supply routes. It was a clever strategy. Without supplies, a military force cannot fight effective battles.

In the Battle of Bloody Ridge, three hills became important. The hills were numbered 983, 940, and 773. The hills were also named Heartbreak Ridge and Bloody Ridge. To the east of the hills was an area called the Punch Bowl. The map below illustrates this area.

THE PUNCHBOWL AREA

Elevations in meters

What made these hills even more challenging was that they were located north of the 38th parallel. UN forces would be fighting inside enemy territory.

South Korean troops were assigned to take Bloody Ridge.

To support their attack, UN Allies opened fire on the area with a heavy artillery barrage. The firepower decimated trees and foliage. The terrain became a barren wasteland. On August 18, the South Korean troops moved in for the main attack. After such a heavy bombardment, the South Korean soldiers felt confident. They moved up the ridge line, assuming only dead bodies remained of the enemy.

But the Communists popped out of their defensive positions—alive!—and opened fire with rifles and machine guns. They also threw deadly hand grenades.

In secret, the Communists had turned this ridge into one big bunker. Trench lines ran across the hills and connected to the heavily fortified protective structures. Most of these bunkers were so well constructed that only a direct hit would destroy them. The largest of the bunkers could hold up to sixty men along with mortars, small artillery, and machine guns. Even worse for the South Koreans, the Communists had littered the terrain with deadly minefields. One wrong step could kill a man.

Korea's many hills were often numbered rather than named.

Intense fighting broke out on the ridge and soon descended into hand-to-hand combat. Finally, after struggling for a week, the South Korean troops finally pushed the Communists off the ridge.

The victory was short-lived.

The very next day, the Communists reinforced their lines and launched a counterattack. Although the US 9th Infantry raced to bolster the South Koreans, these Allied forces could not hold their position—the Communists regained the hill. The South Koreans suffered about 1,000 casualties, wounded or dead.

Clearly, the frontal attack plan was not working.

Allied leaders developed a new strategy. First, the UN

would expand its front lines of attack. That shift would force the Communists to spread out their own lines, weakening them. The plan's second part was a diversionary battle that would trick the enemy into moving away from what actually was the main battle. That diversionary battle would take place in the Punch Bowl.

Two miles wide and surrounded by mountains, the Punch Bowl was filled with bunkers that were covered by heavy timbers, rocks, and dirt.

On August 31, the two engagements began. The US Army launched a hillside assault to spread out the lines, while another contingent struck the Punch Bowl as a diversion.

Recoilless rifle in action in Korea.

Under supporting fire, US troops moved up the hill. Four machine guns blazed away, pinning down the enemy. Unfortunately, a cloak of fog also moved up the hill, forcing the machine gunners to pause the cover fire. It was too risky to fire ahead blindly into the mist when their own men were also moving up the hill. The men had only walkie-talkies for communication, which worked only sporadically. With no observers and no way to see what lay ahead, the men felt almost blind.

And in the pause of their cover fire, the enemy leaped out of the hillside defenses. They rolled grenades downhill. The explosions halted any effective advance.

To regain momentum, reinforcements were sent in, including three BAR men—nicknamed for their Browning Automatic Rifles. But the enemy immediately cut down the first BAR man. Although the other two BAR men kept firing into the enemy bunkers, the structures held.

Shelling a hillside in Korea.

But persistence paid off.

The men kept attacking—attack after attack after attack—until the bunkers began to crumble. But as soon as these men destroyed one bunker, a fresh burst of firepower would erupt from another bunker. It was like a deadly game of whack-a-mole.

The Battle of Bloody Ridge was a mission of utter persistence—never giving up no matter how bad things looked.

Meanwhile, also on August 31, the diversionary mission launched into the Punch Bowl. Fighting alongside South Korean Marines, the US 7th Marines drew enemies away from the Communists' main line. But the weather was a huge problem. Torrential rain began to fall on the Punch Bowl. Heavy water pounded the ground, drowning out man-to-man communication and turning the earth to mud.

But once again, persistence paid off. The next day, September 1, the Marines charged into the enemy using flamethrowers and grenades. They drove them out and took Hill 924.

In fact, the Marines fought so hard they eventually ran out of ammunition. Instead of pausing the battle, they turned to hand-to-hand combat. Foxhole to foxhole, they attacked the enemy, refusing to quit.

One of the most heroic warriors was Marine Corporal Joseph Vittori.

During one heavy counterattack by the enemy, Vittori charged forward alone. His one-man diversionary attack allowed his struggling company to consolidate and strengthen their positions. But in his charge forward, Vittori was isolated from his fellow Marines. For protection, he took a defensive position in a machine-gun nest. The enemy attacked him, but Vittori did not waver. Single-handedly, he moved among the foxholes, blasting away until the machine guns ran out of ammunition before racing to another machine-gun position.

All the while, his company was pinned down. They could

not help him.

Yet Vittori remained persistent. He refused to give the enemy any ground, even as some of them came within a few feet of his position. Sadly, Vittori was mortally wounded in this brave fight. But his valor saved his entire battalion from collapse.

Soldiers under enemy fire creep over Korean hills.

When his fellow Marines finally reached Vittori's body, the ground was covered with the bodies of more than 200 enemy soldiers.

After weeks of relentless fighting, the Marines finally took control of the Punch Bowl on September 21.

Meanwhile, the men back at the main attack site had kept up their own persistent fight from bunker to bunker. With increased firepower—machine guns, grenades, and BAR men—they began overtaking and clearing the enemy bunkers. But when they moved forward to capture the high ground, they faced three more enemy bunkers, all of which opened fire. The UN forces brought in even more firepower—mortars, bazooka, flamethrowers.

On September 5, the UN forces took the ridge. And waited for another brutal counterattack.

It never came.

The UN forces had lost about 2,700 men in the Battle for Bloody Ridge. But the Communists lost many more—an estimated 15,000. Rather than risk more losses, they chose to flee the area.

Powerful weaponry made a difference in this battle, but the victory belongs to the men who refused to give up despite fighting in terrible weather against extremely tough opponents who hid in bunkers which seemed almost indestructible.

Persistence won the Battle of Bloody Ridge.

WHO FOUGHT

Herbert K. Pilila'au was born and raised on the island of Oahu, Hawaii. He had thirteen siblings—eight brothers and five sisters.

HERBERT K. PILILA'AU was deeply religious. Before being drafted into the Army and sent into the Korean War, Pilila'au had considered registering as a conscientious objector. That's someone who refuses to join the military based on their deeply held religious beliefs.

During the Battle of Bloody Ridge, Private First Class Herbert K. Pilila'au's company held a key position on Heartbreak Ridge. The enemy dispatched wave after wave of soldiers. These near-constant attacks depleted the platoon's ammunition.

Pilila'au's company was ordered to withdraw from the area.

But with constant fire coming from the enemy, withdrawal was extremely dangerous. Someone needed to stay in position and keep up a counterfire on the enemy so the other men could leave the area. Pilila'au volunteered to stay. Alone.

As the platoon retreated, the Communist soldiers charged Pilila'au's position.

Pilila'au fired on them until his automatic weapon ran out of ammunition. Then he threw hand grenades. When he had no more grenades, he used a trench knife and his fists to fight the enemy in hand-to-hand combat. Pilila'au is believed to have killed more than forty enemy soldiers before he died.

He was twenty-two years old.

Private First Class Herbert K. Pilila'au was posthumously awarded the United States Medal of Honor "for conspicuous gallantry and outstanding courage above and beyond the call of duty."

Herbert Pilila'au was the first Hawaiian in US history to receive the Medal of Honor.

BOOKS

Rite of Passage: The Korean War Through the Lens of a Young US Marine
    by US Marine Sgt. Charles Baggett, a first-person account of the
    Battle of Bloody Ridge.

INTERNET

Historic news film of the battle with some terrific live footage: youtube.com/watch?v=a7_SuBHMmwU

MOVIES

War Stories with Oliver North: Hill Battles of Korea

BATTLE OF PORK CHOP HILL

April 16–18, 1953

US fighters with M-46 Patton tank atop Pork Chop Hill.

WE ARE NOW going to fast-forward to 1953.

What had happened during the last eighteen months?

Brave men continued to fight on the battlefields of Korea while politicians and governments met for peace negotiations. On one side of the negotiating table were the Communists, made up of representatives from the Soviet Union, China, and North Korea. On the other side were the UN representatives, most of whom came from the United States.

These two sides argued over many topics, such as boundaries dividing North and South Korea and POW exchanges—when and how would these prisoners of war be returned to their homelands?

The UN representatives took these diplomatic talks serious-ly. Three years of brutal fighting had killed and wounded thousands upon thousands of soldiers and Marines, with no victory in sight. The Chinese, however, used these political negotiations as a ploy. They had no intention of surrendering this war. In fact, they still planned to win it.

Meanwhile, inside canvas military tents in remote regions of mountainous Korea, weary warriors hoped their commander would finally tell them, "We're going home."

But as the peace talks dragged on, Major General David Ruffner employed a plan to take control of twelve strategic hills. Ruffner wanted to set up a strong defensive line that would allow his men to command the high ground above the Chinese forces.

Two of these twelve hills were called Old Baldy and Pork Chop. Both were held by Chinese forces.

Men take cover in ditches as a UN convoy gets ambushed. *Harry S. Truman Library photo*

Initially, the US forces managed to capture eleven of these twelve hills, including Pork Chop.

Old Baldy remained in enemy hands.

But on March 23, 1953, a sudden reversal occurred.

The Chinese went after Pork Chop Hill, raining down some 8,000 rounds of artillery fire before the Chinese ground forces stormed forward.

Two US companies from the Army's 7th Infantry Division were holding Pork Chop.

"We just got into our foxholes on the finger [trench] of Pork Chop," recalled Corporal Joe Scheuber, "when enemy mortar and artillery hit us. To our right, more incoming rounds. Then we saw Chinese behind us and realized we were surrounded. We fell back to the trench line at the top of the hill, but the Chinese had reached it first. Hand-to-hand fighting broke out. There was a tremendous amount of noise. I got nicked in the arm and my helmet got shot off. I worked my way down the hill, killing a Chinese soldier with a grenade."

The Battle of Pork Chop Hill had begun.

It became the longest battle of the Korean War.

Shelling a hillside in Korea.

After this first major attack by the Chinese, the hill grew eerily quiet. But the Chinese were sending in reinforcements for an even more powerful attack on Pork Chop. They also set up loudspeakers and played melodic American love songs, all the favorite tunes from back home. Between songs, a Chinese-accented voice speaking English would say things like, "Don't you miss home?" and "We are going to kill you. Why not surrender?"

The songs and statements were all part of the Chinese strategy of psychological warfare. The Communists wanted to destroy hearts and minds, not just bodies. And they knew these exhausted men had been fighting for three long and deadly years in Korea. The Chinese created a campaign to wipe out any remaining morale—the spirit and discipline—among the UN forces.

Ninety-six men were defending Pork Chop Hill. In order to detect the enemy's position, they set up "listening posts." Men worked their way to the front of the trenches and used radios and runners—guys who literally ran back on foot—to alert the men behind them. Some twenty men went forward to the listening posts. The other defended the crest of the hill from reinforced bunkers inside the trenches. The bunkers were set at 30-yard intervals and reinforced with sandbags and timbers along with improvised roof coverings.

However, this setup created some problems. The men were defending such a large area, it stretched out their forces. If one section of the hill was attacked, the other men might not even realize it.

On the night of April 16, a patrol of fifteen soldiers were sent out. Their goal was to set up an ambush at the bottom of Pork Chop Hill. After several hours of surveillance, they noticed about fifty Chinese soldiers moving in.

Sergeant Henry W. Pidgeon realized his men were heavily outnumbered. Pidgeon threw hand grenades at the enemy and ordered his men to retreat.

Hand grenade thrown at Communist positions in Korea.

To protect Pidgeon and his retreating force from capture as they raced back to the trench line, the soldiers above them from the 7th Infantry's Easy Company fired off mortars. But the Chinese opened fire in the darkness and pushed the US forces deeper into the trenches and bunkers. The Chinese forces raced forward without encountering any counterfire and attacked a platoon of soldiers. Unfortunately, because of the rough and steep terrain on Pork Chop and the stretched-out lines, not everyone knew about this attack.

But First Lieutenant Thomas V. Harrold was listening in the

dark. Harrold was in charge of Easy Company on top of the hill. When he heard the increase in enemy artillery and gunfire, he fired a red star rocket—the signal for being under attack— followed by a red star cluster. That second flare meant "give us flash"—search lights.

The searchlights flashed. Then the US forces could see exactly where the enemy was coming from. They opened fire.

The battle became a duel between warring artilleries.

Imagine sitting in one of these trenches at night. Incoming artillery shells scream across the darkness. And you don't know where the shells will land. Lowering your head, you try to dig deeper into the ground, making yourself the smallest target possible. Any of these blasts might kill you. Metal shards ricochet from the exploding shells. Rocks shoot up from the ground. You glance down the trench. A few yards away, your buddy who was alive just minutes ago is now dead. Panic invades your heart—you could be next.

Bunker blast in Korea.

The artillery duel raged for nearly thirty minutes.

Then everything went quiet.

The men of Easy Company cautiously slipped out of their bunkers—only to find the Chinese all around them.

They were surrounded!

Firefights broke out in the trenches and fox holes.

From his command post inside a reinforced bunker, Lieutenant Harrold and four men secured the hill's rear slope. That stopped the Chinese from coming up that direction. But by then, the Chinese had captured most of Pork Chop Hill. And the US forces had even fewer men to defeat them.

Harrold radioed his situation to Colonel William B. Kern, commander of the 31st infantry. Kern sent reinforcements—one platoon each from Love Company, Fox Company, and Easy Company.

Unfortunately, Fox Company got lost on the way to help.

Love Company, under the command of Lieutenant Earl L. Denton, came within about fifty yards of one bunker when Chinese machine-gun fire opened up and never relented. Denton ordered a retreat. Out of sixty-two men in the original company, only twelve would make it back alive to the command post.

With reinforcements from a new company, King Company, Lieutenant Colonel John N. Davis ordered a counterattack at dawn.

First Lieutenant Joseph G. Clemons, Jr., was commanding King Company's 135 men. They were supposed to come up the hill on Love Company's right flank. Unfortunately, the companies didn't coordinate their movements.

Vast mountain ridges in Korea.

The men of King Company were loaded down with ammunition—three hand grenades and extra ammunition belts for each man. Six men also carried Browning Automatic Rifles with twelve magazines for each gun. Another team carried a light machine gun and five extra boxes of ammunition. Other weapons included a flamethrower and a heavy rocket launcher, such as a bazooka, for each platoon. All that extra weight slowed the men's climb up the mountain—at three in the morning in complete darkness.

But Lieutenant Clemons realized they needed to reach the top of Pork Chop Hill before daylight.

"Hit the hill hard," Clemons told his platoon leaders, "and get to the top as fast as men can go. Success depends on speed. We must close before daylight."

Meanwhile, US artillery had been hammering Pork Chop to force the Chinese to keep their heads down, preventing them from firing on the men climbing the hill.

But even without enemy fire, this climb proved extremely challenging. The men's heavy packs were loaded down with ammunition and weapons, and due to the hill's slippery slope, every step forward triggered a slide backward. It took King Company about thirty minutes just to cover the distance of two football fields.

What the men saw horrified them.

King Company's Sergeant Samuel K. Maxwell later said, "We managed to get over the first line of barbed wire through holes cut by shell fire and by walking on bodies of men laying on the wire to hold it down."

The men confronted the first Chinese trench and threw hand grenades, clearing the enemy soldiers from the ground.

At 4:30 a.m., the US artillery fire stopped.

At 5:00 a.m., the Chinese artillery opened fire.

As the men in King Company labored uphill, they couldn't see Love Company where they were supposed to be on their flank. Not only did Love Company not coordinate the attack with King Company, they came under extreme artillery barrage and retreated.

King Company was on its own.

The King Company squad approached a main enemy bunker. Sergeant Walter Kuzmick moved forward. As Kuzmick raised his arm to throw a grenade, the bunker door flew open—and there stood a person who was not a Chinese enemy. He was a survivor from Easy Company—holding a grenade in his hand, too! Seconds later, three Chinese artillery shells hit the bunker. Several men were wounded in the explosion.

Despite the losses, King Company continued to push ahead through twisting and turning trenches and stumbling upon hidden bunkers, constantly fighting the aggressive and seemingly endless Chinese soldiers.

"Pork Chop was a maze," Sergeant Maxwell said, "rats nests of bunkers, line and camo trenches, shell holes and rock clumps. The Chinese kept feeding fresh troops into their counterattacks. The survivors of the previous attacks would then come out of cover and join them. We fought with the men we had. Every hour, we numbered less."

The maze of hills, trenches, and bunkers amid Pork Chop Hill.

Clemons estimated the Chinese held two-thirds of the hill.

He'd already lost half his men in this fight. And there weren't enough remaining men to take the rest of Pork Chop. He called up the third platoon.

And then—surprise!

Love Company showed up. But of the company's sixty-two men who started up Pork Chop Hill, only twelve survived.

Combining Easy and Love companies, Clemons now had about sixty-five men.

Meanwhile, the Chinese seemed to have an endless supply of soldiers.

Things were going from bad to worse.

But another surprise—suddenly reinforcements appeared.

These were the men of George Company who were led by Clemons' brother-in-law, First Lieutenant Walter B. Russell. Clemons thought his brother-in-law was training in the United States. But there was no time to celebrate their reunion.

The Chinese dispatched more reinforcements.

Clemons had asked headquarters for more water, food, radios, ammunition, and blood plasma for the wounded. But the only supplies to arrive were water and K rations—packages of food.

At 1:00 p.m., First Lieutenant James Blake arrived with orders for George Company to withdraw, and for the wounded of Easy and Fox Company to be sent to the rear.

Clemons realized there was a serious breakdown in communications. Headquarters clearly misunderstood the fight these men were in.

"When [those wounded men] go out," Clemons told headquarters, "it is not reasonable to expect that we can hold the hill."

Headquarters responded with ... silence.

"We must have help," Clemons pleaded, "or we can't hold the hill."

Eighteen men of King Company had been killed. Another seventy-one were wounded. Only twenty-five men remained who were capable of fighting.

Clemons regrouped those men and took the highest point on Pork Chop Hill.

The Chinese then launched a bombardment. It lasted for hours.

At about 4:40 p.m., Clemons sent another desperate message back to headquarters.

"We have about twenty men left still unhurt. If we can't be relieved, we need to be withdrawn."

Finally, his message reached someone who understood the predicament. Major General Arthur G. Trudeau made the decision: "We will hold the hill. Send in reinforcements."

But as the reinforcements arrived to relieve Clemons' force, the Chinese artillery barrage killed nineteen of them.

Finally, the artillery fire stopped, and Clemons began moving his men down the hillside.

Twenty hours prior, Clemons had 135 men.

He left Pork Chop Hill with only seven survivors.

The Battle for Pork Chop Hill raged on for several more months. US forces held the mountain until July 27, 1953, when the warring countries signed an agreement to stop active hostilities. That agreement designated Pork Chop Hill as part of a demilitarized zone—or DMZ—that forbids any military personnel or military activity. This DMZ would now separate North and South Korea.

Army soldier Whitney LeBlanc fought at Pork Chop Hill. He received many commendations for his bravery on the battlefield, including two Bronze Stars. But LeBlanc had mixed feelings about this war—which would later be called the Forgotten War.

"Sometimes, sitting there in the trenches with bullets whizzing around and people getting killed," he said, "I would wonder what this was all about. Who and what were we fighting the war for. It wasn't for our country or our independence."

The Battle of Pork Chop Hill is considered a sad and telling symbol of the Korean War. So many lives lost, for so little gain.

WHO FOUGHT

Joseph G. Clemons

JOSEPH CLEMONS GREW up in Florida and later graduated from the United States Military Academy at West Point.

For his leadership while serving as platoon leader in King Company during the Korean War, Clemons earned the Distinguished Service Cross for "heroism in action against the enemy in the Battle of Triangle Hill. Clemons led multiple counterattacks with limited ammunition, including hand-to-hand fighting."

On April 17, 1953, after being assigned as the commanding officer of King Company while still a first lieutenant, Clemons led that counterattack on Pork Chop Hill. He was awarded the Silver Star for his part in this action.

His citation notes "gallantry, personal example, and outstanding leadership" in taking and holding the position, and that "rarely in combat history has a force of the size committed on Pork Chop taken such losses … and nevertheless continued to hold their position."

Clemons was the main character of *Pork Chop Hill*, a book written by S.L.A. Marshall that was turned into a 1959 movie of the same name.

In the movie, movie star Gregory Peck plays Clemons.

BOOKS

Audio Book: *On Hallowed Ground: The Last Battle for Pork Chop Hill* by Bill McWilliams

INTERNET

Some of the best photos of the battle are collected here:
nl.pinterest.com/l_kempner/pork-chop-hill-korean-war

MOVIES

Pork Chop Hill (1959)

THE FORGOTTEN WAR

Front page of a US newspaper, July 27, 1953.

ON JULY 27, 1953, the Korean War ended—unofficially.

No country declared victory.

No country admitted defeat.

And no official peace treaty was signed.

Instead, this war concluded with an armistice—a mutually agreed upon "cease-fire" signed by representatives of the warring countries.

So what did the Korean War accomplish?

Almost nothing, even after three years of horrific battles.

Moreover, the soldiers and Marines returning to America were shocked to discover almost nobody knew—or even cared—about this war.

Navy Lieutenant Junior Grade James "Tim" Timidaiski flew Corsairs on missions throughout Korea. During one mission his plane caught fire. Timidaiski barely made it back alive to the aircraft carrier. That day, however, was just like any other day fighting this brutal war. A humble man, Timidaiski never expected glory or honor. But upon his return to America, he was stunned by people's lack of knowledge.

"People would say to me, 'Where have you been? I haven't seen you in awhile.' I'd say, 'I've been in Korea.' They'd look at me for a moment, kind of blankly, then say, 'Oh. What's going on over there?' Most people had no idea we were fighting a war."

This lack of knowledge was partly due to President Truman not following the US Constitution. Instead of requesting permission from Congress to go to war in Korea, Truman went to the United Nations instead. But another part of the public's ignorance came from human weariness. The United States had recently endured four terrible years of World War II. Most people didn't want to hear about another military action. And many veterans wanted to avoid discussing the horrors of this war. These brave men simply got on with their lives.

UN soldiers leave Korea—and native Koreans return to their homeland, 1953.

With so little attention paid to it, the Korean War became known as the Forgotten War.

Even decades later—right up to today—most people do not understand what happened in this war.

The people who will never forget, however, are the South Koreans.

Invaded by hostile Communist forces, South Korea could have plunged into darkness if not for the foreign help that drove out the enemy. Firsthand, the South Korean people witnessed the high price paid for their freedom. They also paid the highest price of any country, with nearly one million dead or wounded.

US soldier comforts a grieving infantryman in the Korean War.
US Army/Sergeant 1st Class Al Chang

To memorialize the sacrifices made, South Korea has erected historic monuments that include foreign veterans of the Korean War. The mountain once known as "Hill 235," for example, is now officially named "Gloster Hill" in honor of those brave Scottish soldiers who fought and died on its slopes.

On the other side, the North Korean people remain under a Communist dictatorship. Their current leader is Kim Jong-un. His father was the dictator during the Korean War. Some say that the son is worse than the father. Kim Jong-un keeps an iron fist clasped around his country, controlling every detail, from how much food and water people can have to which jobs they can work. He's known for executing even his closest advisors.

The vast differences between North and South Korea—or Communism and freedom—is visible in the satellite image below. The photo shows a night view of the Korean Peninsula. Above the 38th parallel, North Korea sits in darkness. The country's electricity is controlled by the Communist Party. The only lighted areas are Kim Jong's palaces. Below the 38th parallel, however, South Korea glows like every other modern civilization with democratic elections and economic freedoms.

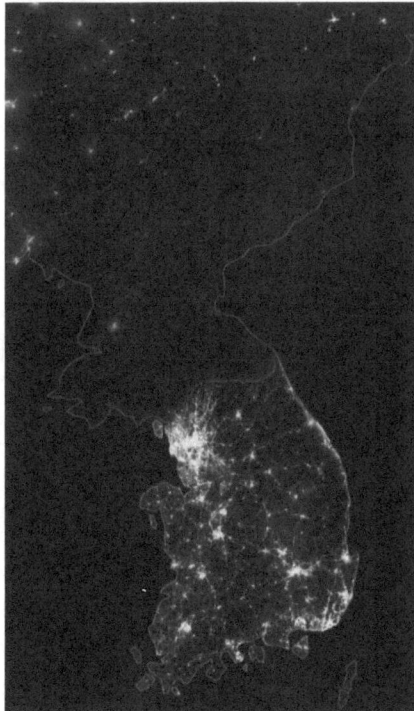

Unlike many people, you now understand what happened in this Forgotten War. If you are fortunate enough to meet a Korean War veteran—many of them are very old now—be sure to thank him for his service. These men fought under brutal conditions. They froze on foreign mountains. They were outnumbered in almost every battle. And they came home to their own country only to find nobody knew about their hardships and sacrifices.

If you ever visit Washington, DC, be sure to visit the Korean War Veterans Memorial.

It was dedicated on July 27, 1995—forty-two years to the day of the war's ending. The memorial marks a shift in public perspective as people realized this war and the men who bravely fought it should never be forgotten.

Korean War Veterans Memorial, Washington, DC.

The memorial's nineteen stainless steel statues look eerily lifelike. The soldiers are dressed in full combat gear, wearing ponchos over their uniforms to stave off Korea's bitter cold. The fourteen Army soldiers, one Air Force pilot, one Navy sailor, and three Marines advance over granite strips and juniper bushes that represent Korea's rugged terrain. A pond of water reflects these nineteen figures—doubling their number to thirty-eight, representing the 38th parallel separating North and South Korea. Nearby, a polished granite wall displays photographic images and lists the names of the veterans, along with this crucial statement: "Freedom Isn't Free."

After reading about this war and these battles, my hope is that *you* now understand why freedom isn't free. And that you never forget the Korean War.

ABOUT THE AUTHOR

JOE GIORELLO GREW up in a large Italian family in Queens, New York. After hearing firsthand stories of relatives who served in World War II and Vietnam, he took an interest in military history. He teaches a popular class for boys called "Great Battles." Joe's goal is to remind young people that "freedom isn't free" and that history is anything but boring.

You can contact him through his website:
www.greatbattlesforboys.com

And on Facebook:
facebook.com/greatbattles

Find all of Joe's books here:
greatbattlesforboys.com/books

Made in United States
Troutdale, OR
12/28/2024